An Effective Strategy for Safe Design in Engineering and Construction

An Effective Strategy for Safe Design in Engineering and Construction

David England & Dr Andy Painting

WILEY Blackwell

Registered Offices
John Wiley & Sons, Inc., 111 River Street, Hoboken, NJ 07030, USA
John Wiley & Sons Ltd, The Atrium, Southern Gate, Chichester, West Sussex, PO19 8SQ, UK

Editorial Office
9600 Garsington Road, Oxford, OX4 2DQ, UK

For details of our global editorial offices, customer services, and more information about Wiley products visit us at www.wiley.com.

Wiley also publishes its books in a variety of electronic formats and by print-on-demand. Some content that appears in standard print versions of this book may not be available in other formats.

Library of Congress Cataloging-in-Publication data
Names: England, David (author), Painting, Andy (author).
Title: An effective strategy for safe design in engineering and
    construction / David England and Andy Painting.
Description: First edition. | Hoboken, NJ : John Wiley & Sons, 2022. |
    Includes bibliographical references and index.
Identifiers: LCCN 2021041749 (print) | LCCN 2021041750 (ebook) | ISBN
    9781119832034 (hardback) | ISBN 9781119832041 (pdf) | ISBN 9781119832058
    (epub) | ISBN 9781119832065 (ebook)
Subjects: LCSH: Industrial safety. | Buildings--Safety measures. |
    Engineering design.
Classification: LCC T55 .E445 2022 (print) | LCC T55 (ebook) | DDC
    658.3/82--dc23
LC record available at https://lccn.loc.gov/2021041749
LC ebook record available at https://lccn.loc.gov/2021041750

Original cover photography and design: © David England
and Dr Andy Painting

Set in 9.5/12.5 STIXTwoText by Integra Software Services Pvt. Ltd, Pondicherry, India
Printed and bound by CPI Group (UK) Ltd, Croydon, CR0 4YY

C9781119832034_120122

Contents

Figures

Tables

Foreword

As a Chartered Surveyor and Fellow of the Chartered Institute of Building, with some fifty years of wide international construction industry experience, I have seen all too often the importance of "getting it right" during the design stage of any project. Ensuring that a project is delivered safely takes planning and cooperation and extends beyond just ensuring the safety of the workers at the construction stage. We saw the value of safety—and safe delivery—at the London 2012 Olympic Park, as well as my own experiences in delivering such notable designs as the concrete ski hill in Finland for the Winter Olympics; the triple water towers in Kuwait City; 55 Lombard Street and Thames House. In 2018 I was appointed to the executive board of the Institute of Construction Management (ICM) and in 2019 I started to construct a digital gateway for CDM professionals to access the exciting and vitally important new world of building information modelling (BIM).

Despite explicit construction regulations having been with us since 1994, we are still witness to waste and error and, of course, appalling tragedy. This book uniquely provides sound, in-depth but straightforward advice at what can only be regarded as one of the most critical stages in the recent timeline of the industry—marking a paradigm shift into new ways of working, thinking, and procuring construction in a fast-evolving, digitally connected world. The message for the reader of this book is just how vitally important it is to effectively manage the improvement of our built environment. The shared expertise contained in this book is so phenomenally timely.

The authors are perfectly positioned to lead in the writing of this book at this critical time in the industry—driving safety, quality, and eliminating error to create important safe projects. This book is important to all those who are interested in construction, engineering, and the built environment, and perfectly demonstrates the duty of care we owe to those who will build, operate, maintain, and perhaps live or work in the things we create.

Since 1994, a prime focus for the industry has been creating a culture of integration, of better safety and workplace health. In those early days I was one of the very first tranche of fewer than two hundred construction professionals to envisage where the future of the safety culture needed to be positioned within the sector. Unfortunately, my early clear vision soon became disillusioned and so, for the last twenty-seven years, I have led teams championing change to construction design and management culture, proudly being recognized in 2008 by the Health and Safety Executive with one of only three awards in the UK as "Health & Safety Champion of The Year." This commitment to safety is something I continue today as head of the ICM Competence Working Group, something that is supported by the ideals and approach contained in the wealth of content in this book. The pragmatic sequencing of safety described in this book I believe profoundly helps to solve how to view the whole landscape and detail of any project, whilst at the same time ensuring the effective management of risk.

I am personally proud and delighted to introduce the unique content in this rather special reference book—enjoy the good read; then read again and reflect.

David F Jones
FCIOB FASI MRICS MIConstM

Introduction

Aims of the Book

Design is the cornerstone of creating and producing any structure, product, or item either for bespoke use or mass reproduction. Anything that is created, constructed, or manufactured relies on design whether for aesthetic, functional, or critical purposes. Of paramount importance is the designer's understanding of the intended use and the application of the product and their subsequent ability to translate this into a finished design. Some examples of products that require specialist design knowledge are:

- Architecture—such as habitable or commercial property or structures.
- Electronics—such as printed circuit boards or electrically controlled devices.
- Marine—such as ships, oil rigs, jetties, and quays.
- Mechanical—such as mechanized plant, engines, and wearable or implantable medical devices.
- Chemical—such as nuclear, biological, and explosible materials, or structures that contain them.
- Emergent technologies—where designers may be dealing with novel production techniques or exotic materials.

Of equal importance to the designer is an understanding of the operating environment in which the product is to be used and how this environment is controlled by such considerations as regulations, standards, or social norms. These considerations may have a direct influence (such as the regulations surrounding health and safety) or indirect influence (such as ethical or moral concerns) on the design process.

Additionally, the actual individuals who will use the product should be considered, as well as any others who may come into contact with it. What is important in any design process is that the criteria of the design requirement are developed within this sphere of considerations and that the product is capable of being subsequently produced accurately to that requirement. This is known as the input-process-output cycle.

This book aims to explain this cycle in detail in order to provide the reader with a broader understanding of the responsibilities of the designer not only to their profession and industry but also of the wider implications of their output by explaining the many considerations that any design should take into account. These considerations are

An Effective Strategy for Safe Design in Engineering and Construction, First Edition.
David England & Dr Andy Painting.
© 2022 John Wiley & Sons Ltd. Published 2022 by John Wiley & Sons Ltd.

not always apparent and it is the product of not only successful designs, but also successful *design management*, that ensures that they are appropriately considered in the design process.

Equally, we aim to demonstrate the important connection between *good* design and *safe* design and how this can be achieved as well as show how the various design professions, with their own standards and practices, are often a reflection of each other, and how design can be improved through the application and management of effective safety.

There are many associations, organizations, and standards, active in a wide range of design disciplines, that aim to improve the design process for either the designer or the client. This book intends to demonstrate that the essential tools for improving safety in design for both designer *and* client are already well established and readily available but, possibly, not well understood. By utilizing these tools any design project can be improved in terms of safety, quality, cost benefit, and project outcome.

Who the Book is For

This book is intended for use by all stakeholders who are involved in the design process, either directly or indirectly, as well as students of any discipline where design is a component part of their syllabus. It is also intended for those who have a responsibility for specifying during the design process and, of course, for those who have an interest in understanding more about the process and the best practice that can be achieved in this demanding and rewarding profession. This book is therefore aimed at:

- designers;
- clients;
- design managers and supervisors;
- those with oversight for design—such as project managers, surveyors, and insurers;
- principal designers (duty holders under the Construction (Design and Management) Regulations 2015);
- specifiers—such as Building Control representatives;
- procurers—such as marketing, sales, or financial departments;
- manufacturers, constructors, and developers;
- students of engineering, architecture, software development, and so forth;
- suppliers.

How the Book is Structured

Different disciplines—or professions—where design is practised tend to generate their own language for the inputs, processes, and outputs that they perform and it is not the intention of this book to attempt to harmonize these differences. Instead, a glossary is provided in this chapter in order that the reader can disseminate the information contained herein and translate it, as required, into the language or phraseology with which they are familiar in their own profession.

Each of the first five chapters deals with a separate component or consideration of the design process. The last chapter prescribes an effective strategy for managing the logical sequence, from the initiating need prior to design commencing, through to the proposed or anticipated disposal of the product. Whilst it is clear that not all design disciplines require all elements of this book's design management process, it is the hope that the reader should become familiar with the *generality* of its intention. The chapters in the book are concerned with:

- the design process—the life cycle of the design process, its influences, and the expectations we have of it;
- regulations—how they affect design and how they can be used for effective process control;
- design management—the tools and techniques used to manage the design process;
- risk—identifying, managing, and controlling all aspects of risk in design;
- design strategy—applying the techniques of safe and successful design.

The final chapter on effective design strategy uses, as a guiding framework, the Construction (Design and Management) Regulations 2015. This is the third incarnation of the United Kingdom's statute interpretation of the European Union temporary and mobile construction site directive 92/57/EEC. The reason for this may not seem immediately obvious to the reader, but we shall demonstrate how the extensive reach of this legislation over the variety of disciplines and objectives to which they apply, combined with the spirit of the regulations with respect to good design management, make them an excellent benchmark.

Although these regulations deal predominately with what may be considered to be the "traditional" construction industry (that is to say, buildings and structures), the spirit of the legislation is to improve the safe function of design in any given project. Regulation is just one of several influences on design and, therefore, to utilize that regulation to the benefit of the design process—rather than consider it burdensome to it—can only provide positive results: by improving management of the design process; ensuring legal compliance; and providing a considered, proper, and safe design output (see Figure I.1).

Construction regulations as a separate piece of legislation were originally introduced in 1994 in response to the high level of injuries and fatalities in the industry historically and they remain one of the many pieces of legislation concerned with workplace health and safety. The third version of the regulations in 2015 encompassed a number of changes which we believe are of fundamental importance not only to the functional requirements of health and safety, but also to the wider moral and financial implications of good design. Moreover, the commercial release of designed products into the UK marketplace is governed by various safety regulations and it is an ambition of this book to encourage the reader to plan their particular project with this knowledge in mind.

Whilst not every design process will require every element described in this book, the reader is encouraged to identify which elements are salient to their particular project or discipline. The establishment of a well-defined and structured environment in which to conduct any design project is a feature of regulations, standards, and practices, which

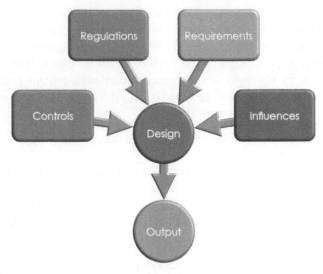

Figure I.1 Design Inputs.

we shall examine in more detail throughout this book. The advantages of creating such an environment, whatever the size or complexity of the project, include:

- better translation of the client's originating requirements for the finished product;
- clear parameters within which the designer can operate;
- control over scope creep or design variance, whether or not intentionally;
- the capture and control of risks which promotes—
  ○ the reduction of errors;
  ○ the control of costs.

Promoting Safe Design

The central tenet for this book is to reinforce the concept of safety as a critical part of the design process. Not just from the perspective of preventing harm to people, but also in ensuring the safety of the project: that is, preventing errors and miscommunication that can cause delays, increase costs, and reduce the quality of the finished product.

There are many factors which can influence any design and these will be explored in detail throughout the book. These factors can have greatly varying effects on a particular design of course, depending on what it is for, how and where it is intended to be used, and how it is to be produced and, if necessary, distributed. There are, though, three factors that affect the scope of all design projects and which are key not only to the designer's appreciation of the fundamentals of the design requirements, but are also an imperative part of the client's decision-making process.

Time, cost, and quality—often thought of as a triangle—affect the scope of any project through being inextricably linked: affecting one will always have an effect on either of the other two or, will affect, either positively or negatively, the quality of the output (see

Figure I.2 Time, Cost, and Quality Balance.

Figure I.2). In this sense, they can be thought of as a see-saw, where the effect of increasing or decreasing any one element can be imagined. What can also be imagined is the necessary response in order to return to equilibrium. Demanding a better-quality product will either affect the cost of it or the time it takes to produce, for example. Demanding a less expensive product may make it quicker to produce but will invariably reduce the quality. Understanding their interrelation and, most importantly, understanding how the client views this interrelation with regard to their requirements, will help to ensure that the design remains as relevant to the client's needs as possible.

The concept of a safe, effective design strategy overall can help to promote the following factors:

- The reduction of harm occurring to users or operators of the product.
- Ease of maintenance methodologies through the thoughtful provision and layout of systems and access points.
- The reduction of losses occurring during the product's life cycle.
- The prevention of errors in the design, which necessitate redesigns and reworking.
- The inclusion of external factors in the product's requirements resulting in a more mature design output.
- The inclusion of expectations from the product's use, and environment of use, which will provide a better experience for the end user.
- A design solution that "learns" from previous examples and builds upon them to advance technologies, techniques, materials, and experiences.
- Better experiences for future users from having a well-considered and well-documented disposal process.

Example Case Studies

Throughout the book, we shall be using five hypothetical case studies in order to highlight the differences between the various concepts and processes discussed. These five have been chosen to demonstrate that, whilst there may be disparity between perhaps the levels or intensities required of each process in comparison, they are all intrinsically linked by the core process of design management. Often this is for different reasons and equally for differing outcomes. The impression that should be gained, however, is that proper design management—and the need to work towards the safest possible outcome—is relevant to *all* projects.

A precis of each study is given below in order to provide background to the reason for choosing them.

Nuclear Power Plant

A nuclear power plant is arguably the zenith of critical design input in terms of operation, maintenance, and disposal. Despite worldwide public concerns, they continue to contribute a large part of non-fossil-fuelled electricity generation in several countries. Everything to do with this type of infrastructure is on a huge scale: preliminary works, design, construction, maintenance, and, of course, disposal, which—in terms of the waste they create—can be counted in thousands of years.

Office Block

Worldwide, the office block has long been a rather dowdy and functionary building. In the late 20th century, however, novel architectural solutions were being developed as a result of new materials being available and the desire of clients to incorporate other spaces into the design; such as accommodation, retail, and leisure. In the aftermath of the global pandemic of 2020, the value of offices as a workspace began to be questioned and once again architects are developing novel ways of enhancing and repurposing these buildings.

Warship

Naval fleets were once populated with many types of specialized vessels: frigates, destroyers, battleships, support ships, and so forth. In recent years the tendency has been towards fewer, large command vessels combined with smaller, lighter vessels which can fulfil a multirole function. Although the operators of warships are highly trained, they often have to work under extreme conditions. Reliability is an absolute requirement. And, as has long been the case, warships often get sold on to other navies after their initial period in service, so the ability to remove sensitive equipment and materials is important.

Home Printer

Printers for domestic use are generally designed on a strict cost/quality basis and with an eye to having relatively short-lived periods in service due to the market forces and upgrades to consumables. Internally they are often composed of proprietary components but externally they must fit the client's aesthetics and brand image. Made in large numbers, design errors can cause large-scale, even potentially worldwide, recall issues.

Motor Car

Ostensibly, the car has changed very little from that developed by Karl Benz in 1885, which was, in turn, a progression of self-powered vehicles that had been developed for over a hundred years before it. Critically built to a price point, the design must take account of aesthetics, aerodynamics, brand image, safety, security, a raft of legislation in each national marketplace, as well as the knowledge that the vehicle will be operated by persons of widely varying levels of skill. Composed of many third-party components, the failure of any of which can cause long-term reputational damage to the client's brand, vehicles can have several operators during their life and their final disposal has become a serious concern environmentally.

The Context of Design

Design and the Product Life Cycle

To understand the design process, we should consider the connection between design and the life cycle of the product which is being designed. The main stages of this life cycle are:

- design;
- production (including manufacture or construction—see Glossary);
- in service (including maintenance and repair);
- disposal (including demolition, repurposing, or recycling).

At each stage of the cycle, a design process may interact with it in a number of ways (see Figure I.3). At the beginning, there will be an initiating need for the product, whether this is a warship, a housing estate, or a software application. The need for the product originates from an entity which may be an organization, an individual, or even another project, which we call the *client*. At this early stage, the design process is concerned with identifying the intention of the design and might (and indeed, should) include radical and unorthodox solutions. The feasibility of producing the various conceptual solutions will be tested, a specification for the product decided upon and, ultimately, a final design will be delivered from which the product may be produced.

During the next stage, the design process may be involved in solving emergent issues and risks that may be encountered during the production process; or it may be necessary to incorporate advances in production techniques or materials. Pre-production proto-types may be required to test certain production techniques or material combinations and the design may be necessarily shaped further by these outcomes. Sometimes very large or complex products can take so long to reach the final commissioning/validation stage that they can be affected by such advances, or even by economic, political, or social events, or upheaval.

When the product reaches the "in-service" stage, there may be such issues as fettling or snagging that require possibly minor design changes. Or, there may be a requirement for substantial changes over time, if the original intent for which the product was designed changes. All the while, the amended design must interact fully with the existing one in order not to compromise the safe use of the product. It is also possible that for mass-produced products, mid-life amendments are possible due to feedback on the product's use, or perhaps due to changes in external factors such as regulations or societal demands.

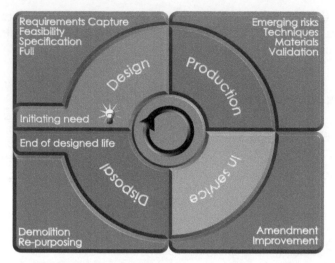

Figure I.3 Product Life Cycle and the Design Process.

Other questions that we must ask during the design process, connected with the product in general, are:

- Is the product feasible?
- Is the product required?
- Can the product be technically produced and in the required quantity?
- Is there a marketplace for the product?
- Are there acceptable limitations of use for the product?
- Is there the capability and capacity to produce the product?

These are fundamental questions that can be regarded as risks associated with the design of the product. Risk is a complex area of analysis that may not be readily understood nor even identified within the design process and we shall cover this subject in more detail later on. For now, it is merely enough to say that an appropriate understanding of risk at the earliest stages of the design of a product is important in providing a safe design solution, in terms of production, use, and disposal.

Upon reaching the end of its designed life, the product will then undergo disposal. This may mean destroying or demolishing it (such as in an office block); repurposing it (such as with a warship); or recycling it (such as with a motor car). Or it may even have been designed to be sustainably managed in a "circular" manner, so that the product's end of life forms very much part of its origination—also known as the circular economy. In any of these events, the design process may need to provide solutions for such things as temporary works or methods of removing contaminants. Additionally, in the case of repurposing or recycling, there may be a need to redesign the product for its new role, hence the circular style of the diagram (see Figure I.3). In any case, it is entirely possible that the original design process at the start of the product's life cycle should have incorporated the intended disposal method within its considerations.

Influences on Design

The design, then, affects a product's use, maintenance, repair, and so forth, but also the use of a product can affect design, either of the product or subsequent versions of the product that follow. The single simplest mechanical device—the wheel—is an example of this. The advantages of using rollers to move loads led to the development of the axle-mounted wheel, which, in turn, saw them being used in a wide variety of applications where they were often redesigned in order to be more efficient. By adapting a wheel by adding a series of pins perpendicular to the rim allowed the user to turn another adjacent wheel synchronously. This, in turn, led to the design of the cogwheel, which is to be seen in practically every mechanized device on earth.

This symbiosis between design and use means that we cannot isolate one from the other (see Figure I.4). A designer cannot design a product without understanding to what purpose it will be put and, similarly, a user or operator cannot use or operate a product without knowing its design limitations. And as safety is the overriding moral and ethical requirement for the design and use of any product, we cannot perform either function—designing or operating—without recourse to the legal, technical, and moral requirements for safety within the parameters of the product that we are designing.

Figure I.4   Product and Design Symbiosis.

Figure I.5 External Influences.

Use, however, is but one part of a product's life cycle; and safety is but one concern for which the designer must find solutions. We can see there are various external influences and demands, including those of design itself, placed on any product for which a solution must be formed (see Figure I.5).

Not all products will of course be influenced similarly by all of these factors but they will all, to some extent, impose considerations or limitations on any design solution. They will also affect the design throughout the life cycle of the product in varying ways too. To what extent will depend very much on the client's originating need, their intention for the product, its operational environment, and the regulatory environment in which it operates.

Let us consider these influences on the product life cycle through these four stages using our five illustrative case studies to which we shall refer to throughout the book (see Table I.1 to Table I.5 inclusively).

Table I.1 Home Printer Influences.

Home Printer	
Design	This is a generic product, widely available and relatively straightforward in purpose. It is unlikely to yield many conceptual variations.
	Many of the printer's components are proprietary. Design will probably be more about corporate aesthetics than radical changes to the layout or function.
Production	Proprietary components are unlikely to require further design changes. Novel processes (such as 3D printing) may require design engagement.
In service	Use of the product may be counted in years. Original design errors may cause the recall of many thousands of units globally. Updates to operating software can lead to malfunctions or obsolescence.
Disposal	Likely to be considered household waste. The use of as much recycled and recyclable material as possible will help to mitigate the waste impact.

Table I.2   Nuclear Power Station Influences.

Nuclear power station	
Design	Despite nuclear power having been available since the 1950s, the design of reactors remains quite fluid due to safety issues and public concern.
	The design of the plant will involve ground studies, environmental issues, reviews of nearby watercourses, seismic studies, wind analysis, and many other surveys prior to even beginning the process of designing a complex structure.
Production	Due to the complexity and length of time to produce the plant, many design issues may arise. There may also be advances in materials and techniques that may need to be incorporated after the final design is established.
In service	Use of the product may be in decades. Original design errors may lead to expensive redesign of complex systems or structures or to inestimable harm and loss.
Disposal	Waste will require specialist handling for many thousands of years. Disposal of the plant itself may take decades depending on the separation of hazardous materials with their surroundings.

Table I.3   Office Block Influences.

Office block	
Design	The need to house a quantity of people in one place to perform administrative work has been around for many years. Recently this has been in combination with living, retail, and relaxation space to increase market value.
	Changes in life/work balances, and the issues of welfare and viral transmission have created opportunities for re-thinking the office workspace and such issues as access, egress, and air circulation.
Production	Although greatly composed of "standard" construction techniques, the ability to increase or decrease the dimensions of the workspace calls for novel building techniques and materials.
In service	Flexibility is becoming a key element of building use as landlords and owners engage with the risks of future usage of their buildings.
Disposal	Buildings with the greatest flexibility in design will be the easiest to transfer ownership of. The use of pre- and post-stressed elements can pose a specific hazard during demolition. Environmental disposal concerns can influence many material choices their the design.

Table I.4 Warship Influences.

Warship	
Design	Modern nations are moving away from large fleets of single purpose ships to more multirole vessels that can fulfil peacetime as well as combat roles.
	Modern weaponry may be more compact than previous types but often requires large amounts of energy to power it. Stealth technologies can have an effect on space and provision for innocuous items like anchors and portals.
Production	Sometimes built in multiple locations simultaneously, modern warships may also use advanced materials alongside regular ones.
In service	Designed to operate for several decades in potentially harsh conditions, warships should be capable of being maintained and repaired whilst under way (i.e., whilst at sea). Multirole vessels must be capable of transitioning from one role to another easily and readily.
Disposal	Warships often transfer ownership at the end of their first life cycle. The removal of weapons systems and hazardous or secret materials or items is vital prior to transfer or scrapping.

Table I.5 Car Influences.

Car	
Design	Under the body, the modern motor car is little changed from its invention in the late 19th century. Power plants and drivetrains are the most likely sources of conceptual progress, with motor sports providing incremental innovations in materials and techniques.
	Heavily influenced by regulations and crash standards, the exterior of cars is also dominated by aerodynamics. Internally, design is influenced by anthropometry and conventions to reduce operator errors.
Production	From vehicles being completely hand-built originally, modern cars are now often laser welded or chemically bonded by robots. The use of any fuel source requires careful location and protection.
In service	Cars may pass through several owners in their lifetime with varying levels of operator skill and possibly little maintenance. Failures of any kind may affect the brand image overall.
Disposal	Environmental concerns now dominate issues such as material reuse. The reuse of old parts may assist older cars to remain on the road but play no part in the brand's image

Preventing Error

The importance of proper design management cannot be underestimated in ensuring the quality of the final output. The design stage is where all considerations for the output must be made in order to prevent errors later on. These errors can be manifested, for example, as: failures to deliver the output on time or in the format originally required; increases in cost; a product which requires essential modification after delivery or where it is unsafe to use in its final state. Where a designed product requires interaction or operation by persons after its delivery, it is vital that this operability also forms part of the considerations of the designer. This is a philosophy which we call "safe to operate and operated safely." This means that a product has been *specifically* designed with the safety of the operator or user in mind and, similarly, that it is *used* safely for the purpose it was designed for. It is about embedding safety into the very core of all products from the very start.

Safety as a Design Component

Safety is a concept with which we are all familiar. To design a product that is not "safe" is unethical from a professional point of view and immoral from a societal point of view. But safety in this regard is not the only context in which we can put the word "safe." We have already briefly touched on some of the factors that can influence a product, its use, and its design. If we take the word safe to mean "without harm *or loss*" then we may apply it to a number of these factors (see Table 1.6).

The invention and introduction of the power loom in England in the late 1700s, and the various improvements made in the early 1800s, saw a meteoric rise in the use of machinery in the textile industry. This of course led to the migration of large swathes of the population from the countryside to the cities and subsequently paved the way for dramatic socio-economic change in Britain throughout the period. The early looms were manually controlled by the weaver who required training and experience in order to operate them properly. The later Lancashire Loom, from around 1840, was semi-automatic and therefore could be operated as part of a number of similar machines by only one person. This person required a smaller skill set than a fully-trained weaver as the machine was conducting many of the complex routines autonomously.

Whilst many of these machines may not have been designed with explicit safety features, it could be said that they were not dangerous to operate per se, given sufficient training. What was dangerous—and the cause of many horrific accidents—was their maintenance. This was often carried out by young children who were able to fit nimbly under the frames of the machines to collect the cotton dust and fluff, which could clog

Table I.6 Application of Safety Across Various Factors.

Operation	Application of safety
Exceptions (faults, failures, and breakdowns)	The provision of complete and accurate information regarding the product is supplied to the end user, allowing them to identify exceptional occurrences and respond appropriately, thereby minimizing production losses.
Maintenance	Repairs and maintenance—both scheduled and exceptional—can be conducted without causing injury or ill-health to the maintainer.
Financial	The design accurately reflects the client's requirements thereby preventing increased costs due to redesign or rework.
Project management	The design project is delivered on time and within budget due to accurate communication and cooperation between all project stakeholders, thereby preventing financial loss.
Amendments	In-service design changes are performed accurately with full consideration of the existing product's capabilities and limitations thereby preventing unnecessary production losses.
Disposal	Residual risks inherent in the product are accurately recorded for the end user allowing end-of-life disposal of the product without harm to the persons disposing of it.

the machine, or cause the risk of fire—and all whilst the machines were running. In addition to the risk of being maimed or killed, there were ill-health effects such as breathing in dust and hearing disorders from working alongside noisy machinery. We can see from this example that even if a design may be considered safe in use (for a given value of safe!), other external influences such as maintenance or repair may lead to harm or loss. The likelihood and the severity of these losses occurring is the risk; but risk—or risky—is not the opposite of safe.

We examine risk and its management in a later chapter; for now, we shall merely look at how to reflect risk in the examples (see Table I.6). For the risk of some harm or loss occurring in one of the situations given, there must be some *exclusion* or *omission* to have occurred. For example, under "operation" we have concluded that the "use of the product does not cause injury or affect the health of the user in the normal parameters of operation." For harm or loss to occur, therefore, we must have excluded or omitted some part of the design process that ensures the safe operation of the product. Or, alternatively, the use of the product falls outside of the normal parameters of operation; that is, someone uses it in a way, or in a place, *for which it was not intended*. In either case, the likelihood of this happening, and the severity of the outcome should it do so, are entirely dependent on the product, its use, its environment, and the persons operating it.

All of these factors should be known—or at least considered—by the designer. And whilst the designer may have absolute control over their ability to prevent errors and omissions in their design detail, they may have little or no control at all over how the product is used, maintained, or disposed of. The essence of safe design is, therefore, to *manage the risk* of these uncertainties to reduce, to the lowest possible or acceptable level, the possibility of harm or loss occurring throughout the product's life. This is done through the introduction of various control measures which will have an order of precedence—or hierarchy—that determines how effective they are.

Introduction—Summary

The process of design is dependent on a large number of factors, over and above the actual requirement of the client—or driving force—behind the need for the design in the first place. And whilst it may be readily admitted that not all designers will be faced with all of the aspects of design that we discuss in this book, it should be incumbent on all designers to be cognizant of *all the potential criteria* that affect their profession. Recognizing the wider implication, limitations, and impacts of one's profession is an effective way to further one's own competence and the design profession as a whole.

We have discussed the critical importance of safety in the design process, and this will be discussed in more detail throughout the book. Safety, as we shall demonstrate, reflects far more than preventing harm to persons. Safety, and its antonyms threat and danger, can be applied to any number of matters—financial, reputational, environmental, societal—throughout the life of the designed product and we shall examine the tools that are available to the designer to enable them to assess, control, and mitigate these.

We have also discussed the core constraints of time, cost, and quality that apply to every design project, and we have looked at how these interact with one another like a balance or see-saw. These three constraints must always be considered in connection with safety—if cost is reduced, for example, will safety be affected? Moreover, we must also understand the end result of the design process—the product itself. Where is it to be used, and by whom? Will it need to be maintained and if so, by whom and how? What is the marketplace for the product and what restrictions or regulations might apply? Has a previous similar product had issues or are there advances in materials or techniques that we can deploy—or at least advise the client of—that will improve the product's performance, longevity, safety, or reputation?

And finally, we must be aware of the designer's critical role in the development and production of everything that we use in our world today. Whatever it is that we design, our input is crucial and it is our responsibility to ensure that we have produced the safest possible solution within the remit of the design specification.

Glossary of Terms

Client The originator of a requirement for a **product** and the ultimate authority of its specification. A **client** may be an individual, a corporate body, or a requirement from another **design** process created from some arising need.

Construction The act of building or making a thing. Whilst usually associated with the traditional view of house building and so forth, this term is used in this book to refer to any **product** that has to be physically manufactured. See **production**.

Core constraints The de facto considerations in all projects of time, cost, and quality; the manipulation of any one of which will always affect the other two, either positively or negatively. (Decreasing the amount of time available for the project, for example, will either increase the cost or decrease the quality of the **product**.)

Cost-benefit analysis The systematic process of assessing the cost of developing a particular **design** or **product** against the perceived benefits that it would bring about.

Design The intentional creation of a **product** to a given specification to fulfil a specific purpose.

Design freeze The point in the **design process** where the full technical design is established and the **design** is prepared for the **production** stage. Changes to the **design** beyond this point, unless in response to a **design risk**, should be considered to be outside of the designer's scope and contract.

Design intent The overarching reason for a **design**. An aircraft that can land on water, or accounts software that covers all financial aspects of a closed organization are examples of **design intent**.

Design life cycle The period during which any **product** is subject to the **design** process. This ends when the **product** is accepted into service. Further isolated **design** life cycles may occur throughout the **product**'s life: motor cars, for example, are subject to continual improvement throughout their production and use.

Design management The structured and competent control of the **design** stage, which ensures the systematic analysis of each phase, provides clearly documented evidence, and allows full and free cooperation and communication between all relevant **stakeholders**.

Design risk assessment An assessment of the **risks** associated with the **production**, operation, maintenance, and disposal of the **product**. Used to demonstrate the mitigation or control measures that have been implemented in the **design** process in consideration of these **risks**.

Error The act of omission in human behaviour. Forgetting to add up a column of numbers, making a mistake in converting units of measure, or pressing the wrong button on a machine due to work pressure are all examples of errors.

Full design intent The agreed detailed **design** that results in a **design freeze**. The final **design** from which **production** can take place.

Hazard Something with the potential to cause harm or loss. Examples are: electricity, machinery, transport, chemicals, pathogens, investment, information.

Practicable That which can be put into practice. Note that this is as opposed to "practical," which implies something is reasonable or serviceable.

Produced See **production**.

Product The output of a **design** and **production** process. **Product** in this book is used as an overarching term to cover anything that has to be **designed** and created, by whatever means.

Production The act of physically creating a **product** from raw materials, which includes intelligence (i.e., that which produces a piece of software, for example). This term refers to any of the following: **construction**; manufacture; assembly.

Project An undertaking to create a **product**. From the originating idea, or requirement, to the final handover of the completed **product** to the **client** or end user.

Relevance The importance to the **client**, on a sliding scale, of the critical requirements of a design concept. Can be used for **cost-benefit analysis.**

Risk The possible outcome of a **hazard**, usually measured in terms of likelihood and severity. Examples are: electricity can cause a shock; chemicals can cause health issues; information can cause operational failure (when misinterpreted or recorded incorrectly).

Safety Case A structured evidence-based argument that justifies the reasoning that something is safe to use for a specified purpose within a specified operating environment.

Stakeholders Anyone with a vested interest in the **product**. From the board room to the shop floor operatives, from those who design it to those who use it, there may be many people or groups who have an interest in the **product**.

Structure Any physical thing that has been **constructed**, manufactured, or assembled, including: buildings, machines, means of transport, excavation.

Validation The process of proving that a **product** has been **produced** fully in accordance with the **design** criteria.

1

Elements of the Design Process

The design process consists of several stages that commence once it is agreed that the idea for a product is worth the investment of time and resources. To establish this prior to starting the design process in earnest, the following are required:

- an initiating need, or driving force, that requires resolution;
- a business case to justify the investment.

Initiating Need

This pre-design stage is concerned with understanding the high-level goal or ambition of the client as well as what this is intended to resolve. For example, the goal may be a new design of car in order to seize upon a current gap in the marketplace, but car design obviously encompasses a great many styles, types, performance levels, and design possibilities. Refining that goal during this stage to "a car that can carry five persons comfortably, with luggage, and be powered electrically" reduces the number of possible design outputs for consideration and may also reduce the overall design costs.

Some possible reasons and underlying factors for an initiating need might include those in the following list.

- To support current growth in the business through:
  - adding an extension to existing premises;
  - adding new premises, either acquired or purpose-built;
  - moving en masse to a new location;
  - developing an additional or enhanced range of products for the market.
- To support a change in technology or invention that will:
  - increase production;
  - reduce operational costs;
  - improve safety and reduce levels of risk.
- As a result of new or changed legislation that:
  - is designed to improve air quality;
  - is designed to enhance water quality;
  - impacts the use of restricted or hazardous materials;
  - makes material changes to workplace assets.

An Effective Strategy for Safe Design in Engineering and Construction, First Edition.
David England & Dr Andy Painting.
© 2022 John Wiley & Sons Ltd. Published 2022 by John Wiley & Sons Ltd.

- To effect a change to the client's undertaking due to:
  - a recent safety-related case that highlighted poor practice;
  - an expansion into new markets or product lines;
  - preparing for the sale of the undertaking;
  - a change in direction of the board or a fundamental injection of capital.

The initiating need may lead to a number of initial ideas that could push the limits of technical capability and possibility. Whilst some of these ideas might be ruled out very quickly, some of the more novel or unorthodox ones may help the client re-evaluate their ambition for the product. It may transpire that for a modest capital investment, the client can blaze a new trail in their particular sector, perhaps in aesthetics, technology, or product sustainability. This balance of investment against return is resolved through conducting a business case.

Business Case

This stage is where the outline costs and benefits of progressing with the initiating idea are captured. There may well be more than one possible solution to any requirement, each of which may have different investment costs (capital expenditure) and running costs throughout its life (operational expenditure). Combined with this will be the level of risk each option exposes the business to. The business case must balance these positive and negative outcomes for each option against the levels of exposure to risk—in terms of financial, reputational, and corporate risk, and so forth—that the organization is prepared to accept. This should provide the client with a balanced argument from which to make an informed decision. There is also always the "do nothing" option to consider if this is ascertained to be less impactive to the organization than the value of the investment in an ambitious new project. Each organization has a level of risk it is prepared to accept under various circumstances and this in itself will play some part in the potential design options considered later on.

The business case of course provides an informed approval based on knowledge at the time it was conducted. Should issues arise during subsequent stages of the design process, or risks become apparent that had not been hitherto identified, then there should always remain the option to halt the project; either temporarily in order to reassess the business case or permanently to prevent unnecessary further expenditure.

Requirements Capture

Once the initiating need has been defined and the business case to proceed has been approved, a statement of requirements (SoR) is created to document what the client *actually* needs. This is because the client's ambition may include a "wish list" of qualities for the product that cannot be justified in terms of current technology, capability, or probity. The involvement of a wider group of stakeholders, such as suppliers, manufacturers, or teams within the client's organization, should result in a more balanced and informed SoR. This should then be checked against the initiating need to ensure that the key objectives of it are being met.

At this stage, a risk register (covered in the risk management chapter) is also created and populated with risks to the project. Sometimes with complex projects, it may be that the design requirements have to mature before some risks become apparent. This is in part due to how risk can be either foreseeable or observable; that is, foreseeable risks are those that we can speculate with some certainty *might happen* and observable risks are those that *readily present themselves to us*. For example, the requirement might state that the product is intended for use in atmospheres that contain explosible or volatile substances. In this case, the *foreseeable* risk is that the product might cause an explosion if it is not made to the correct tolerances. The *observable* risk is in the capability of the production stage to maintain those tolerances against the design specification.

The statement of requirements should state the client's design intention clearly but without specificity. In our "car" example from the initiating need stage earlier, we might demand electric power with a range of at least 300 km in the SoR. If we were to qualify this by stating that the power is to come from lithium-ion batteries we might exclude, at the project's expense, emerging battery technology; just as lithium-ion supplanted nickel metal-hydride batteries, which in turn supplanted nickel-cadmium batteries before that. The SoR should pose the question to which the full design is the answer, thus allowing the designer some degree of flexibility in proposing solutions.

The Design Process

Once the SoR is completed, the designers can be approached and the design process can begin. This process is composed of four key stages that follow the product's life cycle through to completion of production. These stages may be combined or repeated according to the size and complexity of the design itself (see Figure 1.1). The stages are:

- feasibility;
- specification;
- full (or technical) design;
- validation.

There may also be further independent reiterations of the design process during the in-service stage of the product's life cycle—for example, where the product is repurposed or amended—and these should be considered as separate iterations of the design process.

Design Feasibility

By interpreting the client's needs from the SoR as well as considering the initial risks captured in the risk register, the designer may arrive at the most likely solution quickly. Indeed, the solution may be quite obvious. This should not, however, preclude the designer from suggesting alternatives, regardless of their apparent obscurity. Similarly, the client would do well to examine each suggestion dispassionately in case any of

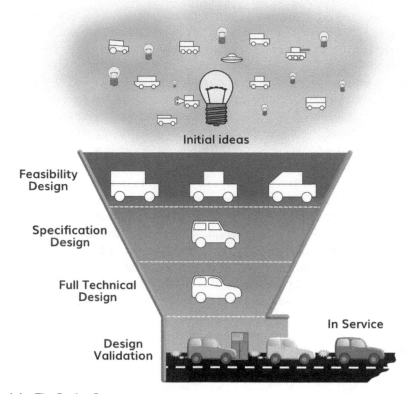

Figure 1.1 The Design Process.

them expose a hitherto unconsidered solution or potential cost saving, perhaps over the lifetime of the product. Such considerations are a matter for the client to review as often as necessary to whittle down the prospective design concepts to those that best fit their business case.

The SoR will assist in reducing the number of initial ideas to a select few that fall within the agreed levels of suitability and affordability. All the while, the designer should be referring to any control measures necessary to ensure the safety of each design (see the section on the general principles of prevention) and to establish that, at this early stage, the risk with each one has been—or is being—considered.

During the feasibility phase, various tests, experiments, and calculations may be required to substantiate one design over another. Once again, the output of these tests should be considered with reference to the SoR and the risk register to ensure that they are providing relevant data.

Towards the end of this stage, a design review might be undertaken to qualify that the SoR has been adhered to and whether the proposed design meets with the client's objectives for the project. The design review will also examine key stakeholder requirements, as well as identify any potential risks concerning design, production, in-service use, and disposal, and these can help to inform the design risk assessment as well as the design specification.

Design Specification

The specification stage is where a single design concept from the feasibility stage is agreed for further development and progression. The design can now be imbued with enough detail in order to fully understand its functionality and any potential impacts associated with the end product. This level of detail may still be a little way from being enough to go to production, but should be enough to provide modelling data and to understand whether it will perform as specified in the SoR.

As the design matures during this stage it will become more complex and this can lead to a tendency to drift away from the original idea. To help prevent this, it will need to be tested against the SoR as well as additional issues such as: whether the product can still be technically produced, whether it creates additional hazards, and whether risks have been captured and managed by the designer. All the while throughout this stage the design should continue to be developed in accordance with the general principles of prevention.

As with the previous stage, a design review may be undertaken to assess whether the design meets with the client's intentions as captured in the SoR and whether the benefits continue to outweigh any perceived risks. The review should question how the design is to be tested to ensure that it is compliant and will meet all the required operational parameters. At this point the design is formally accepted by the client and any subsequent substantive changes made by them will result in any additional costs being borne by the client.

Full or Technical Design

The full or technical design part of the process is where the design gains ever more detail in preparation for production. The design is modified in accordance with all aspects of the stakeholders' requirements, as far as the client demands. It must consider how the product will be produced as well as the implications of time and cost for all relevant testing and certification required in order to demonstrate that it meets the initiating need and the SoR, as well as any relevant legislation and standards. At this stage the designer should be cognizant of any appropriate testing requirements and any operational or functional tests that need to be demonstrated to the client. This will inform the validation criteria required at the end of the production stage and how this is to be managed by the production programme. These criteria are what the producer will use to demonstrate that the product has been completed strictly in accordance with the full design. Where the product is designed to be produced in phases, the validation criteria will also be used to ensure each successive phase begins only after appropriate testing and acceptance of the previous phase is complete.

The full design should continue to be developed in accordance with the general principles of prevention and this will be reflected in the design risk assessment which should be updated as appropriate throughout this stage. This ensures that due consideration is paid to the later areas of in-service use (specifically operation, maintenance, and disposal) and any potential risks during these stages which might be mitigated by amending the full design.

A design review at this stage will consider all these various aspects and provide an agreed "freeze" of the design, in preparation for elevating to the production phase.

Production Phase

During the production phase the "frozen" design is of course the de facto design that the producer will work to. There may, however, be a requirement to alter the design as issues and interactions are identified. Whilst these minor updates always demand the attention of all stakeholders, thereby allowing inclusive and demonstrable decisions to be made on any amendment, there is often also a need to provide some level of autonomy to the producer and allow the installing engineers to make working decisions. This level of responsibility and accountability will have been formally agreed and documented prior to the production phase. Any changes made to the design, irrespective of size, will be captured on updated design drawings. Whether the original designer or another progresses the design through the production phase, they should be aware of how a small change can have a major impact on any subsequent phases, especially in complex projects. Minor changes may also introduce additional risks into the final product, or impose previously undocumented limitations of use.

As production progresses, there may be a need to test or validate elements of the product prior to subsequent elements being produced. Here, the test will be against the element's agreed design and should be in accordance with criteria previously identified in the validation plan. Often in long or complex projects where there is a gradual delivery of the product, the client may become responsible for the maintenance and upkeep of completed elements now under their control. Again, this is subject to formal agreement prior to the production phase and any required documentary evidence from the client to prove the criteria of any maintenance has been met, thereby ensuring any warranties with the final product delivery remain valid.

Validating the Design

The validation design review takes place at the end of the production phase when all the predetermined tests and checks have passed the guidance and tolerances set out in the validation plan. A typical validation plan should contain:

- a brief description of the project;
- a list of the elements that require testing;
- manufacturer/supplier details;
- any unique item codes, references or asset numbers;

- who is accountable for conducting each test;
- who is accountable for witnessing each test;
- what information should be officially recorded as part of each test;
- any items of special test equipment that are required (some may require pre-booking);
- specific hazards associated with each test;
- risk assessments for each test (high energy systems, power systems, working at height, confined spaces, etc.).

The plan should identify all the tests that need to be undertaken, when each is planned, any prerequisites that need to be in place, and what needs to be tested in a specific order (this will identify the critical path of the testing process). There should be signature boxes available to allow both those conducting and accepting the tests to sign them off, as well as any witnesses if required. If applicable, serial numbers and calibration certificates for test equipment used should be captured at the time to prove they were in date during the test. All of this information should be included in the health and safety file, technical file, and/or safety case.

Once the validation plan has been created it should be signed by all stakeholders involved in the test, thus confirming their accountabilities. The plan should then be formally authorized by the client and the conducting authority.

The validation plan can be examined as part of the validation design review to ensure that everything has been installed and is operating as expected and, if there are any arising issues or risks, that suitable mitigations have been put in place.

The review also examines the accuracy and completeness of any maintenance and operational information that has been supplied by the producer to the client. This should also combine with a review of the project's risk register to determine that any residual actions of the production stage have been achieved, or are at least scheduled to take place. These actions may include any that are connected to residual risks with the final product where, perhaps, the mitigation lies in operational or maintenance procedures, or even training regimes, that the client (or end user) must implement prior to the product being used for the first time.

The validation design review may also be used to fully agree the contents of any technical documentation regarding the product that may be required by appropriate regulation. Known as either a "technical file" or a "health and safety file," it takes a specific form of documentation that, usually, has already been laid down during the design and production phases. As this can be of a regulatory nature, the client may wish to seek expert assistance with its creation. This forms part of the "golden thread" of information that helps to inform the owners, operators, and users of products about its safety—in terms of use, maintenance, and repair.

For complex designs such as nuclear powerplants or warships, this validation phase may well take several years, whereby some responsibility is handed over to the client for the day-to-day running and maintenance, with certain items left in the control of the producer until they are validated. The Royal Navy aircraft carriers, HMS Queen Elizabeth and HMS Prince of Wales, were launched and sent to sea years before final acceptance of the flight equipment/systems into operational service.

Often, after all production and reviews have been completed, the senior stakeholders, designer(s), and producer(s) will attend a "lessons learnt" exercise in order to capture any issues with the project and the design. These exercises can prove invaluable in providing feedback to all parties on where the project went well and where it did not. In cases where the client has further projects to undertake, a review such as this can be extremely useful.

Lessons Learned

A valuable component in virtually any type of design process is the conducting of a "lessons learned" review towards the end of the project. Also known as a "wash-up meeting" or "post mortem," this type of review can, when conducted properly, provide valuable information for all stakeholders, particularly for any future projects and, for professional stakeholders such as designers, useful feedback in terms of competency and communication. This can be important in respect of their continuous professional development.

Discussing what went well and what did not go well across a number of elements in a project can often identify sometimes simple but effective changes that can be made in the future. A client not organizing a communication plan sufficiently well; a designer not being provided with a robust statement of requirements; pre-construction information not being supplied early enough—these are all examples of where small detail changes can be made in future projects that will have a positive impact on costs and may reduce potential delays. In effect, learning lessons from previous projects is a strong path to preventing error.

Learning lessons in this way can sometimes be a disquieting procedure—possibly one reason why they are often overlooked. Some stakeholders may view them as merely a way to apportion blame, but this is not the raison d'être. Any project can suffer from hindrances and pitfalls, even when all those concerned are working to a well-defined system to prevent errors. The reason for reviewing the lessons of any project is to attempt to *prevent the same mistakes happening again*; and, by doing so, incrementally adapt each successive project in the pursuit of continuous improvement.

Towards the end of a project, a "lessons learned" document should be sent to each stakeholder or group of stakeholders to elicit from them their responses to the two simple questions: "what went well" and "what did not go well." Responses can be anonymized if need be but this can detract from rooting out problems that may have only been apparent to one type of stakeholder. Difficulties in the supply chain, for example, may be of enormous concern to the producer but the client may not have been aware of them, or even have been unduly concerned if they were. Physically meeting as a group on the return of these submissions will then greatly assist everyone in understanding not only the difficulties that each other had during the project, but also *how* those difficulties were perceived and what they meant in real terms to the project's efficacy.

The elements of the project to be reviewed in this way fall variously under the categories of control, competency, communication, and cooperation and might be arranged under the following topic headings:

- Technical competence. How well did each stakeholder, particularly those in a profession or trade, act during the project? Did anyone deal magnificently with a particular issue or was anyone felt to need improvement in their competency?

- Project management. Was the project managed well? Was information passed in a precise and timely way? Did everyone have the right access to the project at the right time?
- Resource management. Usage of time is important, as is the availability of other resources, such as money, materials, and equipment.
- Organizational management. However well-planned a project is there is likely to be some facet, however small, that can cause disruption to any otherwise smooth progress. Perhaps the introduction at a late stage of an additional stakeholder that had been unforeseen—could this have been better planned at the outset as a "potential risk"?
- Relationship management. Communication and cooperation between various groups can always suffer from disagreements. It is not that these take place but how they are dealt with and reconciled that matters. Was the structure in place to deal with these or did certain groups excel in conjointly dealing with a particular issue?
- Communications. The type, frequency, and output of communications should be planned at the beginning of any project, but did it actually all go to plan, or were there issues that could have gone better or have benefited from more time allocated to them for discussion?

To ensure the success of any review of the lessons from a project, there should be clear definition of its style and framework—as well as its ground rules and ambitions—stated in the communications plan laid out at the beginning.

The Design Process—Summary

The design process begins once the statement of requirements has been agreed and goes on to encompass the feasibility review of the initial ideas all the way through to the delivery of a completed product to the end user. The design is matured, incrementally, from the single design agreed at the feasibility stage through modelling and defining of the specification, to the validation of the final product. Throughout the process the designer must ensure that the design fully realizes, where physically and technically possible, the intentions that the client originally had.

Far from being an obstruction to the design process, scrutinizing the design at predetermined points prevents it from moving too far from the original intention, and thereby saving costs associated with redesign, rework, and failed products. This scrutiny can be as an ongoing procedure or as a set-piece review at the end of each design stage. However it is achieved, any design will benefit from being scrutinized from the outset because the greatest savings—and the best chances of reducing error—are to be gained before the design has matured. Asking questions should never be seen as obstructive.

The original intention of the client—whether they are an organization, an individual, or even another design project—should be firmly understood, not only from the technical aspect of the intended product but also from the initiating need of the client: the reason why the product is required in the first place. Understanding this helps to ensure the relevant stakeholders are involved in the process as well as helping to deliver an appropriately worded statement of requirements that will guide the design process along the right path. And along with the influences of those who will have to make, operate, and maintain the product, we must add the environment in which the product will be used.

Throughout the design process, the general principles of prevention will enable us to ensure that the design matures safely. Only by embedding safety as a *critical component* of the process can we maintain our duty of care to all those who will interact with the product as well as our duties under relevant regulations. Again, this is no obstacle to our endeavours, but a beneficial course of action to prevent error and waste, be that in terms of time, money, or materials.

2

The Regulatory Environment

The Importance of Regulation in Design

In the introduction, we discussed how "safe" can be used to reflect a number of criteria affecting a designed output: operational, financial, maintenance, and so forth. We understand that safe means "without loss or harm" in this respect, and that these influences can be mitigated during the design process. In this chapter, the word safe—and the concept of safety—are very firmly in the context of personal safety; that is to say, the health, safety, and welfare of persons who come into contact, whether by intention or circumstance, with the designed product. This is due to the prevalence of health and safety legislation globally and particularly in the United Kingdom and Europe. This prevalence has seen a marked fall in the number of workplace injuries over the last few decades in many markets across the world and especially in the United Kingdom, which has an enviable record of workplace safety.

Health and safety cannot be ignored as a topic, nor as an environment in which all design must be undertaken. For this reason, it is requisite that a designer has an understanding of the requirements of safety legislation to ensure that, throughout all stages of a product's life cycle, the safety of all those who come into contact with the product are prevented from being harmed. We shall see in this chapter that, although health and safety legislation is a very broad—and yet sometimes very specialized—subject, its application in design can be extremely straightforward and beneficial. Used correctly, some health and safety legislation can also prove to be incredibly useful in delivering well-controlled projects in a safe and timely manner.

For virtually all designed products there will be some form of regulatory or governing requirement or influence for which the designer must account. These may be in the form of:

- industry-specific requirements—such as building regulations;
- national requirements—such as the Health and Safety at Work etc. Act 1974;
- trading requirements—such as the European Community's "CE" marking;
- international requirements—such as those of the American National Standards Institute.

An Effective Strategy for Safe Design in Engineering and Construction, First Edition.
David England & Dr Andy Painting.
© 2022 John Wiley & Sons Ltd. Published 2022 by John Wiley & Sons Ltd.

Some regulatory requirements refer specifically to the design of the product insofar as making it safe for use. Building regulations are just such an example where, for instance, the maximum distances that persons have to travel to a place of safety in the event of a fire are clearly identified for different types of buildings.

Regulations for the use or supply of products, however, outnumber those for their design. The designer should be at least cognizant of the plethora of regulations that apply to the product during the production, commissioning, and in-service stages to help ensure that the design contributes the least obstruction to workplace regulations. In so doing, the designer becomes part of the whole safety life cycle of the product—the concept we call "safe to operate, and operated safely."

Additionally, there may be relevant standards which, although not compulsory, may be required by the client or the client's location or type of business: for example, the Ministry of Defence's DefStans in the United Kingdom and the equivalent MilSpec for the US Department of Defense. In exceptional circumstances, there may be no existing regulations or standards to abide by where the design is in a new or emerging sector. The creation of early telephones at the turn of the 20th century; the first space flights of the 1960s; and the creation of commercial computer software in the 1970s are all examples.

The design of components is driven by, and for, influences that extend beyond the design process. These influences are, successively, those of the immediate user; the corporate body or group supporting that user; the industry to which that body or group belongs; and the society in which that industry exists. This creates a two-way dialogue where society can affect design—through the demand for better safety after a catastrophic event, for example—and conversely where design can change the perceptions and eventual demands of society (see Figure 2.1). An example of the former is the widespread installation of the Train Protection and Warning System on the UK's railways from early 2000 after a number of train crashes, and in particular the Ladbroke Grove rail crash of October 1999 which killed 31 people. Public demand for safer railways prompted a slew of safety measures, both trackside and fitted to locomotives, which has seen a considerable and welcome reduction of fatalities and injuries in the two decades since.

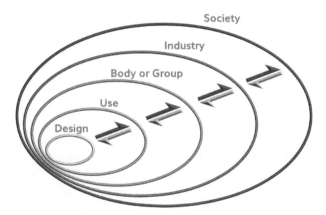

Figure 2.1 Influences on Design and Vice Versa.

The latter situation, of design impacting on society, can be illustrated by the introduction of the Apple iPhone in June 2007. Quite radical in its departure from the accepted design and functionality of mobile telephones of the day, it is hard to imagine today any mobile handset not resembling the iPhone in looks and operation. The way of working for the majority of people is now firmly entrenched in the use of "apps" and the mobile handset has become, for many, a centre point of connectivity and creativity that was perhaps not envisaged only a decade previously.

Where there is no exacting regulation or standard for the designer to adhere to—and it should be appreciated these will be rare events—they must follow the moral code of the society in which they operate. Different societies have differing tolerances and levels of acceptance for the concept of risk and danger. These may be driven by historical, political, or hereditary reasons.

The range of governing regulation is necessary not only to promote a particular benefit in society, such as safety, but also to create a level playing field across all of the spheres of influence that we have discussed. Hence, there are regulations for society, various industries, the use of certain products and so on, as well as those that span several markets, such as European Union Directives. By establishing society's and industry's expectations in law these various regulations may define what we can call legal and not legal, but they can also help us to frame our perspective on design by understanding the reasons behind such regulation: by appreciating the *spirit* as well as the *letter* of the law.

United Kingdom safety legislation was, up until 1974, fairly prescriptive in its approach. The problem with this approach is that as times and technology move on, the regulations are at risk of becoming redundant. It is also sometimes easy to circumscribe prescriptive law for those who have a mind to do so. Today, instead, we have safety legislation that is centred more around the principle of risk identification and reduction being performed by those that actually create the risk; that is, designers, employers, manufacturers, and so forth. This type of approach places the responsibility firmly in their hands to control that risk—and, therefore, any potential harm—to a suitable level. That suitability depends of course on such things as the potential severity of the harm that could be caused, in terms of the level of harm and the numbers of people it could affect, as well as the type of industry or product of concern.

Health and Safety at Work etc. Act 1974

In the United Kingdom, the fundamental requirement for safe design is Section 6 of the Health and Safety at Work etc. Act 1974 (HSE, 1974). To understand this fundamental requirement, we shall now examine the Act's requirements in the relevant clauses of Section 6.

Firstly, it is important to recognize that the Health and Safety at Work etc. Act 1974 is concerned with work-related regulation (paragraph (a) (see Table 2.1)). Therefore, a product being designed for the public market only—for example, a kitchen kettle—would not fall under the jurisdiction of this Act. It would, however, come under the obligations of the European Community's "CE" marking scheme, or that of the UK's

Table 2.1 HASAWA Section 6 (1).

(a) It shall be the duty of any person who designs, manufactures, imports or supplies any article for use at work –

(b) to ensure, so far as is reasonably practicable, that the article is so designed and constructed as to be safe and without risks to health when properly used;

(c) to carry out or arrange for the carrying out of such testing and examination as may be necessary for the performance of the duty imposed on him by the preceding paragraph;

(d) to take such steps as are necessary to secure that there will be available in connection with the use of the article at work adequate information about the use for which it is designed and has been tested, and about any conditions necessary to ensure that, when put to that use, it will be safe and without risks to health.

Conformity Assessed scheme which both impose similar requirements. Secondly, the first line refers to "any article for use at work," which suggests this only applies to the *things* we use at work, not the *places* in which we conduct our work. In this instance, with regard to the structures and buildings we occupy at work, the Construction (Design and Management) Regulations 2015 (CDM) (HSE, 2015) would apply and these, again, follow a broadly similar pattern of requirements to the Act. The reason for this is that the Health and Safety at Work etc. Act 1974 is an "enabling Act," which means that, by virtue of Section 15 of the Act, further subordinate regulations can be made, of which the Construction (Design and Management) Regulations 2015 is but one example.

Paragraph (b), the requirement that articles are "safe and without risks to health" is qualified by two caveats; firstly, that this is "so far as is reasonably practicable." This is a common expression used in health and safety legislation and guidance and stems from the court case Edwards v National Coal Board (1949) 1 All ER 743 (CA) (Edwards v National Coal Board, 1949). The judge in this case stated that "reasonably practicable" was a narrower term than "physically possible" in that a computation must be made where the quantum of risk is balanced against the necessary cost in terms of time, money, and effort. "Reasonably practicable," therefore, requires the individual to demonstrate that if steps to reduce risk have not been taken it is due to the fact that the cost of implementing those steps is grossly disproportionate to the anticipated reduction of risk. It is an important feature of health and safety considerations and the subject of much discussion and contention. It is, however, relatively straightforward to appreciate: it would not be "reasonable" to spend £1 million on measures to protect one individual from very minor injury, but it most certainly would be to prevent serious harm, or worse, to a much larger number of people.

The second caveat is that the article is safe *when properly used*. Clearly, the proper use of something can only occur when clear and adequate instructions and training for its use have been issued with it.

Paragraph (c) requires testing and examination to be carried out "as may be necessary" and this is also a requirement of CE marking, where applicable, as well as being inferred in CDM Regulation 9, "Duties of designers." It is noted that this regulation is also applicable to persons specifying materials for the project who should, it can be assumed, be reliant on the provision of information from the manufacturers about the quality and suitability of those materials for the purposes for which they are intended. The manufacturer in turn would be required to have conducted their own design processes and tests by the same regulations listed here.

Paragraph (d), which requires "adequate information" can also be found in the CE marking regime through the provision of instructional information and a technical file. CDM also requires this in the form of the health and safety file provided at the end of the project. Additionally, this file includes information about any residual risks with the output of the project and this is mirrored in paragraph (c) of Section 6 (1) where it refers to "any conditions necessary to ensure that, when put to that use, it will be safe and without risks to health." The health and safety file should contain, as we shall discover, information on such things as the safe methods of maintaining and cleaning the product.

Section 6 (2) (see Table 2.2) has broad implications for the design of all articles provided for use "at work" and the removal, again "as far as reasonably practicable," of risks to health and safety. This requirement is mirrored by CDM in the requirement for the same elimination or minimization of risk for construction projects (what is considered "construction" by these regulations is discussed later in this chapter).

Risk mitigation can be implemented through the use of the *general principles of prevention*, which are discussed in more depth in the risk management chapter. The principles are a hierarchy of risk control measures, introduced in the Management of Health and Safety at Work Regulations 1999, and in the guidance to CDM they are precised as "eliminate, reduce, control." It is the legal and moral duty of the designer to eliminate risks in the design of any product or, where the risk cannot be eliminated, to reduce or control it by other means, which must then be informed to the user/operator in order that the prescribed methodology of safe operation is laid out.

Table 2.2 HASAWA Section 6 (2).

(a) It shall be the duty of any person who undertakes the design or manufacture of any article for use at work to carry out or arrange for the carrying out of any necessary research with a view to the discovery and, so far as is reasonably practicable, the elimination or minimisation of any risks to health or safety to which the design or article may give rise.

Section 6 (3) (see Table 2.3) places requirements on those who install articles at work—that is to say, contractors. It can be inferred that, for a person to install or erect some article correctly, the provisions of the previous subsections must all have been applied; that is, the correct implementation of the design, the proper testing for the circumstances, the provision of adequate information regarding how the article is to be installed or erected, and the provision of information regarding any residual risks in its use, operation, or function.

Table 2.3   HASAWA Section 6 (3).

(a) It shall be the duty of any person who erects or installs any article for use at work in any premises where that article is to be used by persons at work to ensure, so far as is reasonably practicable, that nothing about the way in which it is erected or installed makes it unsafe or a risk to health when properly used.

Section 6 (6) (see Table 2.4) reinforces what we have said about the specification of articles and materials as part of the design process, insofar as it is not necessary to re-examine and test them. What is important to remember is that the specifier of articles and materials must satisfy themselves of the adequacy of the information supplied regarding any such articles or materials, as well as the competence of the supplier or manufacturer.

Table 2.4   HASAWA Section 6 (6).

(a) Nothing in the preceding provisions of this section shall be taken to require a person to repeat any testing, examination or research which has been carried out otherwise than by him or at his instance, in so far as it is reasonable for him to rely on the results thereof for the purposes of those provisions.

Environmental Protection Act 1990

The Environmental Protection Act 1990 (DEFRA, 1990) succeeded the Control of Pollution Act 1974 and Section 5 of the Health and Safety at Work etc. Act 1974. It established legal responsibilities for pollution control for land, air, and water, as well as statutory nuisances such as noise or smells. It provides a stricter regulatory system over those prescribed in the previous pieces of legislation on controlling environmental emissions. The Act also covers waste management on land, by introducing formal regulation of waste handling, storage, and treatment. This is achieved through waste management licensing which demonstrates an organization's duty of care in the transfer of waste.

Nuisances from commercial or industrial premises such as noise, smells, light, and even infestations are covered by the Act where they are prejudicial to the health and welfare of persons nearby. This includes construction sites.

Construction (Design and Management) Regulations 2015 (CDM)

These regulations, first coming into force in 1994, are the United Kingdom's interpretation of the framework European Directive 92/57/EEC, also known as the Temporary and Mobile Construction Site Directive. Specific construction regulations outside of the various Factories Acts were originally introduced in 1961 due to the poor record of

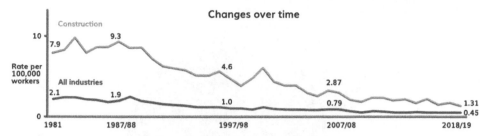

Figure 2.2   Fatal Injury Rate 1981 to 2018–2019 (HSE, 2019).

accidents in the construction industry. In 1964–1965 there were 276 fatalities and 513,000 injuries in the construction industry whereas, in 2018–2019, there were 30 fatalities and 54,000 reported injuries, clearly demonstrating the benefit of individu- ally regulating this industry. There have been dramatic improvements in reducing fatalities in the construction industry in comparison with all industries since 1981 (see Figure 2.2).

The regulations, however, are not centred solely on the health and safety of the physical act of construction. The "design and management" components of the regula- tions are equally important in the consideration of elevating safety as an intrinsic concept in the design process and this is why there are specific duties imposed in this regard on those involved in any project covered by the regulations. The specified duty holders are:

- the client;
- the designer(s);
- the contractor(s);
- the principal designer (where one is appointed).

A 2004 report by the National Audit Office suggested that up to 60% of fatal accidents in construction were attributable to decisions made prior to building work commencing, including during the design phase, clearly demonstrating the hugely important role of design in the implementation of safety throughout a product life cycle; that is, in its production and operational stages.

With similar regard to safety as a concept of preventing loss as well as harm, a report first published by the Get It Right Initiative in November 2015 (GIRI, 2019) sug- gested that as much as 21% of the cost of construction is wasted through error. These errors are manifested through design and management issues that cause direct and indirect financial loss as well as increased wastage (see Figure 2.3). In the United Kingdom, GIRI identified that this cost is in excess of £20 billion per annum. Clearly, this should be reason enough to improve the design process—from both the safety and cost implications during this process—but it may not be immediately clear of the importance of CDM to the design process outside of the traditional construction industry. To understand this, we must first turn to the definition of "construction" in the regulations.

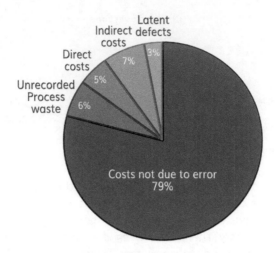

Figure 2.3 Costs Due to Error in Construction (GIRI).

CDM defines "construction work" as "the carrying out of any building, civil engineering or engineering construction work" and goes on to list the activities within those spheres (see Table 2.5).

It can be seen that this list, when taken into account with the prevailing inclusion of "building, civil engineering and engineering construction," perhaps includes more in terms of designed products than it excludes.

It is also pertinent at this point to examine what the regulations define as a "structure" (see Table 2.6).

Table 2.5 CDM Definition of "Construction Work."

(a) the construction, alteration, conversion, fitting out, commissioning, renovation, repair, upkeep, redecoration or other maintenance (including cleaning which involves the use of water or an abrasive at high pressure, or the use of corrosive or toxic substances), de-commissioning, demolition or dismantling of a structure;

(a) the preparation for an intended structure, including site clearance, exploration, investigation (but not site survey) and excavation (but not pre-construction archaeological investigations), and the clearance or preparation of the site or structure for use or occupation at its conclusion;

(a) the assembly on site of prefabricated elements to form a structure or the disassembly on site of the prefabricated elements which, immediately before such disassembly, formed a structure;

(a) the removal of a structure, or of any product or waste resulting from demolition or dismantling of a structure, or from disassembly of prefabricated elements which immediately before such disassembly formed such a structure;

(a) the installation, commissioning, maintenance, repair or removal of mechanical, electrical, gas, compressed air, hydraulic, telecommunications, computer or similar services which are normally fixed within or to a structure.

Table 2.6 CDM Definition of a "Structure".

(a) any building, timber, masonry, metal or reinforced concrete structure, railway line or siding, tramway line, dock, harbour, inland navigation, tunnel, shaft, bridge, viaduct, waterworks, reservoir, pipe or pipeline, cable, aqueduct, sewer, sewage works, gasholder, road, airfield, sea defence works, river works, drainage works, earthworks, lagoon, dam, wall, caisson, mast, tower, pylon, underground tank, earth retaining structure or structure designed to preserve or alter any natural feature, and fixed plant;
(a) any structure similar to anything specified in paragraph (a);
(a) any formwork, falsework, scaffold or other structure designed or used to provide support or means of access during construction work, and any reference to a structure includes part of a structure.

Additionally, there is a schedule to the regulations that defines specific types of activity in any project that must receive particular attention during the construction stage. It should be conferred, therefore, that these activities should also receive particular attention at the design stage, should they form part of the final design intent. These activities are selected for the peculiar risks that they pose or because they are subject to additional health and safety regulation. It should be noted that this schedule applies to activities *during the construction stage*, not during the operational stage. The designer should be mindful, therefore, to consider this during the specification of the design to evaluate if any part of the design causes any of these activities to be adopted in the production process. If so, it is incumbent on the designer—as well as the producer—to consider any mitigation or control measures that may be appropriate (see Table 2.7).

Table 2.7 CDM2015 Schedule 3 Work Involving Particular Risks.

1. *Work which puts workers at risk of burial under earthfalls, engulfment in swampland or falling from a height, where the risk is particularly aggravated by the nature of the work or processes used or by the environment at the place of work or site.*
2. *Work which puts workers at risk from chemical or biological substances constituting a particular danger to the safety or health of workers or involving a legal requirement for health monitoring.*
3. *Work with ionizing radiation requiring the designation of controlled or supervised areas under Regulation 16 of the Ionising Radiations Regulations 1999.*
4. *Work near high voltage power lines.*
5. *Work exposing workers to the risk of drowning.*
6. *Work on wells, underground earthworks and tunnels.*
7. *Work carried out by divers having a system of air supply.*
8. *Work carried out by workers in caissons with a compressed air atmosphere.*
9. *Work involving the use of explosives.*
10. *Work involving the assembly or dismantling of heavy prefabricated components.*

Whilst it is not the designer's place to always exactly specify how a producer may produce a particular product, it should be within their professional remit to have an understanding of the process. As we have discussed, the early adoption of communication between the various parties involved in any project can lead to the better control of risks that may not have been made fully known to all.

Our concern in this book, however, is principally with design and the design process, so we must examine the relevant requirements for these in the regulations. The central tenet to the regulations, in managing the design stage for everything covered under its broad definition of construction, is to firstly promote design that considers the general principles of prevention and, secondly, to provide oversight of the design stage. The regulations provide clear indication of what design is, and this includes:

- drawings;
- design details;
- specifications;
- bills of quantity;
- calculations prepared for the purpose of design.

It may be presumed, then, that a "designer" is anyone who produces, or is engaged with, any of these activities. Again, the regulations provide clear indication of who a designer might include. They are:

- architects;
- architectural technologists;
- consulting engineers;
- quantity surveyors;
- interior designers;
- temporary works engineers;
- chartered surveyors;
- technicians;
- or anyone who specifies, or alters, a design.

The last item in this list is of particular interest due to its wide-ranging potential. A client, for example (and in this context we mean a person or organization), who specifies any element of the design, or instructs the designer to use a particular material or production technique, becomes a designer, and is therefore required to comply with the regulations. As the guidance to the regulations state: "The person who selects products for use in construction is a designer and must take account of health and safety issues arising from their use."

The general principles of prevention are referenced in Appendix 1 of the regulations and are intended to be implemented throughout the construction project, including the design phase. Introduced in the Management of Health and Safety at Work Regulations 1999, they are a guiding principle for all areas of risk prevention throughout health and safety at work. We shall explain more about the principles in the management of risk section but for the time being will simply list them as they appear in the regulations (see Table 2.8).

Table 2.8   General Principles of Prevention in Regulations.

(a) avoid risks;
(a) evaluate the risks which cannot be avoided;
(a) combat the risks at source;
(a) adapt the work to the individual, especially regarding the design of workplaces, the choice of work equipment and the choice of working and production methods, with a view, in particular, to alleviating monotonous work, work at a predetermined work rate and to reducing their effect on health;
(a) adapt to technical progress;
(a) replace the dangerous by the non-dangerous or the less dangerous;
(a) develop a coherent overall prevention policy which covers technology, organisation of work, working conditions, social relationships and the influence of factors relating to the working environment;
(a) give collective protective measures priority over individual protective measures; and
(a)   give appropriate instructions to employees.

The moral requirement for safe design has fomented over many years due to public reaction to disastrous events. Sometimes this may be seen as an immediate response to an event, such as the Grenfell Tower fire, or it may be a more gradual, subliminal process such as the improvements over many decades to electrical safety. However the path to public expectation is formed, it is a fact that today the public mood is far more attuned to safety and welfare than perhaps it has ever been, and unsafe events can have long-term adverse reputational consequences in addition to more immediate safety concerns. Our ethical reasons for providing safe design solutions therefore stem not only from the responsibility we have to our peers and our profession, but also to our client, the contractors we work with, the individuals who will use and operate our designs, and just as importantly, the wider public at large.

In summary, the principles are to avoid risks where possible; evaluate those risks that cannot be avoided; and put in place proportionate measures that control them at source. As previously noted, in the guidance to the construction regulations, they are precised as:

Eliminate – Reduce – Control

The oversight of the design process has been an intrinsic requirement of the construction regulations since their inception, with the duty being performed by a functionary who has had varying titles. In the original 1994 regulations the duty holder was known as the "supervising designer" and this became the "CDM Coordinator" in the revised 2007 regulations. Currently the duty is now the responsibility of the "principal designer" in the 2015 incarnation of the regulations. Despite the title, the principal designer is not responsible for the design per se, but rather the robust management of the design process as well as the important process of implementing the "four Cs": control, cooperation, communication, and competence.

Already with large or complex projects there is often a design manager whose job it is to lead the design team. This position, to be effective in fulfilling all of the aspects of robust design so far considered in this book, must be occupied by an individual with all of the necessary skills, knowledge, experience and, moreover, the independence and facility to critically question the various outputs. It may be considered that if the design manager is employed by the design organization or by the client themselves, this independence might be frustrated by other considerations or obstructions to their impartiality.

Provision and Use of Work Equipment Regulations 1998

The Provision and Use of Work Equipment Regulations 1998 are the United Kingdom's implementation of the European Union's Amending Directive 95/63/EC to the Use of Work Equipment Directive and may not, in isolation, appear to concern the design process. However, simple understanding of some of the fundamental requirements of the regulations can help to ensure that any design output (of a relevant product) can assist the owner/operator to fulfil those requirements when it enters service. It may also help greatly with providing an understanding of the safety environment of the product as well as with the process required for CE marking. Very broadly, the Provision and Use of Work Equipment Regulations 1998 cover the safety of work equipment with regard to a number of features (see Table 2.9).

Table 2.9   PUWER 1998 Applicability of Regulations.

(a) The suitability of the equipment for the purpose intended.
(b) The maintenance and inspection procedures and their frequencies.
(c) Any specific information, instruction and training that is required.
(d) Specific risks, dangerous parts and protection devices.
(e) The controls and control systems, including emergency stop and isolation devices.
(f) Visual elements such as lighting, markings and warnings.

The suitability of the equipment for the purpose intended will be decided at the feasibility stage and throughout the design process. Ensuring there is a robust statement of requirements will give this design phase the well-defined parameters within which to operate and, likewise, a well-informed pre-construction information document will provide detail of the environment in which the equipment is to be used.

The maintenance and inspection procedures, periodicities, and any specific risks, dangerous parts, and protection devices will have been considered throughout the design process and judged against the general principles of prevention. That is to say that any risk from dangerous parts or functions of the equipment should have been reduced, to as low as reasonably practicable, to ensure the safety of the operator, and the maintenance procedures should be able to be carried out without risk of harm. It is often the case that where maintenance or inspection procedures are carried out, they must be done with the equipment in a condition outside of the normal operating parameters: for example, with safety guards removed or interlocks defeated. Designing-in measures that mitigate the potential harm during all exceptional functions of the equipment should be as important as those during normal operation.

Any specific information, instruction, and training that is required would, in the event of a project under CDM, be contained in the health and safety file. Early cooperation with the end users will allow the creation of specific training material to be developed concurrently with the design and production stages or, at the very least, provide the end user with sufficient notification of the type and style that such material might take.

The suitability of the equipment for the purpose intended and the maintenance and inspection procedures and frequencies might also form part of a validation process at the end of a project under CDM, ensuring that the output has achieved its intended aim and that it is capable of being maintained as such.

CE Marking

The intent of CE marking is the harmonization of product quality across the European Union as part of the New Legislative Framework. This framework includes essential or other legal requirements; product standards; standards and rules for the competence of conformity assessment bodies as well as for accreditation; standards for quality management; conformity assessment procedures; CE marking; accreditation policy; and, lately, market surveillance policy including the control of products from third countries.

Although not directly impinging on the design process, the CE marking regime requires—as an output—technical documentation (often referred to as a technical file), which will contain information about the design of the product. Despite being referred to as a "file" the information does not necessarily have to be contained within a single dossier. The "technical file" may be an index of where the relevant information is stored, possibly in noncontiguous locations. The important consideration is that the relevant information is accurate, up to date, and stored in a readily accessible way, and in an accessible format. Considering how the design process is to be recorded and documented at the earliest stage can assist in the creation of the technical documentation later on. Each piece of EU legislation requiring CE marking establishes its own requirements for inclusion in the technical documentation but broadly the requirements are similar (see Table 2.10).

Table 2.10   Items for Inclusion in the Technical File.

(a) *A general description of the product.*

(a) *An overall drawing of the product, as well as any specific drawings such as circuits diagrams.*

(a) *Full detailed drawings, with any calculations, test results, etc., which provide proof of the product's conformity with the essential health and safety requirements of the applicable directive or legislation.*

(a) *Copy of the risk assessment, which has documented the identification and mitigation of applicable risks and hazards.*

(a) *A list of standards and other technical documentation that have been applied as part of the conformity process.*

(a) *Copies of conformity documentation for critical components of the product.*

(a) *Copies of technical reports detailing any assessments carried out.*

(a) *A copy of instructions and other information supplied for the safe use of the product.*

(a) *A copy of the manufacturer's Declaration of Conformity.*

The European Union legislation on the requirements and implementation of CE marking are complex and are beyond the remit of this book. This extract from the Official Journal of the European Union (OJEU)—a Commission Notice on the implementation of European Union product rules and referred to as the "blue guide"—provides a list of the legislation to which CE marking refers (see Table 2.11).

Table 2.11   List of European Union Directives Subject to CE Marking.

Subject	Directive number
The restriction of the use of certain hazardous substances in electrical and electronic equipment	2011/65/EU
Appliances burning gaseous fuels	2009/142/EC
Ecodesign requirements for energy-related products	2009/125/EC
Simple pressure vessels	2009/105/EC and 2014/29/EU
Toys' safety	2009/48/EC
Electrical equipment designed for use within certain voltage limits	2006/95/EC and 2014/35/EU
Machinery	2006/42/EC
Electromagnetic compatibility	2004/108/EC and 2014/30/EU
Measuring instruments	2004/22/EC and 2014/32/EU
Non-automatic weighing instruments	2009/23/EC and 2014/31/EU
Cableway installations designed to carry persons	2000/9/EC
Radio equipment and telecommunications terminal equipment	1999/5/EC and 2014/53/EU
Active implantable medical devices	90/385/EEC
Medical devices	93/42/EEC
In vitro diagnostic medical devices	98/79/EC

(Continued)

Table 2.11 (Continued)

Subject	Directive number
Pressure equipment	97/23/EC and 2014/68/EU
Transportable Pressure equipment	2010/35/EU
Aerosol Dispensers	75/324/EEC as amended
Lifts	95/16/EC and 2014/33/EU
Recreational craft	94/25/EC and 2013/53/EU
Equipment and protective systems intended for use in potentially explosive atmospheres	94/9/EC and 2014/34/EU
Explosives for civil uses	93/15/EEC and 2014/28/EU
Pyrotechnics	2013/29/EU
Regulation on the Labelling of Tyres	Regulation (EC) No 1222/2009
Personal protective equipment	89/686/EEC
Marine equipment	96/98/EC and 2014/90/EU
Noise emission in the environment by equipment for use outdoors	2000/14/EC
Emissions from non-road mobile machinery	97/68/EC as amended
Energy labelling	Directive 2010/30/EU

This list, however, is not exhaustive and the marking of products with "CE" is required by other pieces of European Union legislation not listed here; for example, the Construction Products Regulation 305/2011/EU, which requires the product to be tested against a harmonized standard prior to receiving its CE mark. It may be clear to the reader that this whole area of legislation will potentially require expert guidance in order to fully comply. What should also be clear, however, is that following a defined and predetermined design process will improve both (a) the quality of the documentation required for the technical file, and (b) the timeliness of its delivery to the appropriate body.

Building Information Modelling

The term "building model" and the concept for three-dimensional design and modelling of buildings has been around for several decades. The term "building information modelling (BIM)" was, however, introduced in a white paper by Autodesk (Autodesk Building Industry Solutions, 2002), where they stated that BIM solutions should have three core characteristics, which were to:

- create and operate on digital databases to facilitate collaboration;
- manage change throughout those databases so that any change to any part of a database would be coordinated in all other parts;
- capture and preserve information for reuse by additional industry-specific applications.

The engineering sector has, of course, been modelling and prototyping designs using software applications for many years, including being able to produce both accurate final drawings and bills of materials for final production.

The ability to present and visualize building design in three dimensions can only be regarded as being extremely useful in showing how changes either solve issues or create further issues in surrounding areas. This is especially important for areas of operation or specific maintenance tasks where the location of access panels or switchgear, for example, can be awkward or obstructed. Another benefit is to be able to accurately determine the quantities of materials and fittings required in order to minimize waste and to help promote more efficient delivery scheduling to site. BIM is intended to manage and intelligently use the data collected throughout the product life cycle of plan, design, build, and operate. It is mandatory for new building and infrastructure projects in several countries including the United Kingdom.

Having a system for recording all information concerning the design, production, and safe operation of a building can only be seen as a fundamentally essential thing to do in order to produce a safe design and to understand how that design interacts with users and operators throughout its life. It may also be seen that having functional safety information fully supports the requirements of CDM in the production of a health and safety file, as well as going a long way in support of a safety case. It also supports the idea of the "Golden Thread" of information defined in the "*Building a Safer Future— Proposals for reform of the building safety regulatory system report*" (Ministry of Housing, Communities and Local Government, 2018). This calls for appropriate safety information to be made readily available for all user or owners of building structures and was a direct result of an enquiry into the Grenfell Tower fire in 2017.

Standards

For virtually every design there will be one or more standards that the designer will find applicable to either the design itself, the materials intended to be used, components that need to be procured, or the proposed manufacturing, construction, or installation techniques. These standards may be divided into the following three types:

- Standards that affect the design—for example, Part B (fire safety) of the Building Regulations, which dictate maximum distances to fire exits.
- Standards that affect the specification—for example, BS EN 1994 (Eurocode 4) for composite steel and concrete structures.
- Standards that affect the design process or the management thereof—for example, ISO 9001:2015 or BS 7000.

Standards that affect the design are beyond the remit of this book but it is nonetheless worthy to recognize the influence they have. The designer in any given sphere of their profession should be aware of relevant standards that directly affect design as part of their competency, an element of the four Cs which we will examine in more detail later in this chapter. Standards that influence other areas, such as specification and management, can usually be safely left to other professionals to administer. Some, such as ISO 9001, require a considerable amount of working knowledge in order to apply them correctly. Material specifications should be properly understood by those in the supply chain, again as part of their competency and in line with their duties under Section 6 of the Health and Safety at Work Act.

As part of any discussion around competency, design professionals should be interrogating their own competency for the task for which they are commissioned. Only through frank introspection can the individual truly assess whether they have the ability to provide the appropriate level of integrity to the task, or whether they require some form of support in areas where they are lacking the appropriate skills and experience, or whether they should recuse themselves from the task completely.

Standards that affect the specification—be it of materials, techniques, or peripheral products required in the design—may be myriad, and beyond the comprehension of the individual designer. Here, the designer will be reliant on the information proffered by the manufacturer or supplier and, therefore, there is a requirement to evaluate the quality of that information—in effect, a competency check on the information and the informant. It is imperative that the materials or products are checked for compatibility with other adjacent items being used in the project or, indeed, perhaps already existing in the area where the design output is to be placed or integrated. Similarly, materials being specified perhaps due to price or environmental considerations should be confirmed appropriate for any installation techniques used during construction or special operational requirements upon completion. Installation techniques chosen for their speed, for example, may be incompatible with materials chosen for their quality of finish.

With the specification of materials there is still the opportunity to de-risk the procedure by applying the general principles of prevention. Specifying one supplier, for example, whose products have all been tested interdependently, will mitigate the risk of materials being inappropriate to be used together. This may, of course, affect the time–cost–quality balance in that there may be less expensive products on the market but the *time* that it would take to confirm that they are compatible with one another will add *cost* to the project in any event. Again, it is the provision of a well-defined statement of requirements that will serve the procurement process well, which will, in turn, allow the designer to concentrate on the design process itself and provide the most appropriate output without the risk of variations during the design phase caused by changes of material specification.

There are a number of management standards that may be applied to any project with varying effects on the day-to-day functionality of the design process. BS 7000 for Design Management Systems, for example, is concerned with establishing the tools and environment in which design management in a project can progress. There are a number of subsets to the standard including: managing innovation; managing the design of manufactured products; managing design in construction; and managing inclusive design. In essence, these points are in connection with the design environment, the establishment of which is crucial to the proper and effective management of the design process.

Another standard concerned with the environment in which work function takes place is ISO 9001:2015. Often considered a quality standard, ISO 9001:2015 is usually speci-

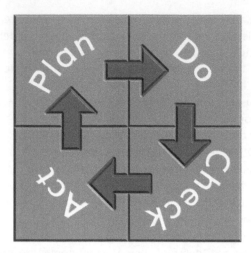

Figure 2.4   Plan-Do-Check-Act Cycle.

fied to demonstrate that an organization has a robust commitment to the control of its administrative function. ISO 9001:2015 is not prescriptive; instead, it requires that not only have a number of considerations concerning the organization and its administrative functions been defined and documented but also that this process is managed, adhered to, reviewed, and adapted as necessary. It requires, at its core level, clear demonstration of the Plan-Do-Check-Act cycle. This is similar to the Plan-Do-Study-Act process (The W. Edwards Deming Institute, 2020), also known as the Deming cycle. Dr. Deming himself preferred the use of "study" rather than "check" to encourage assessment of the results of the plan, rather than merely "checking" its success or failure. The point is perhaps a little moot but nonetheless noteworthy (see Figure 2.4).

The standard does not prescribe the type or methodology of the systems an organization uses to confirm its management of these processes; rather, it requires the organization to demonstrate the robust repeatability of the system and the engagement with it from shop floor to boardroom. The practical upshot of this standard in connection with the designer's work will most likely be in the format and frequency of documents that they produce, along with the inclusion of supporting documents. This will have a significant benefit to the creation of a well-defined project structure in the early stages of the design process, which can be referred to later in the project in the case of misunderstandings, errors, or scope creep.

The "Four Cs"

The "four Cs" are laid out in HSG65 Managing for health and safety (HSE, 2013), the guidance document for the Management of Health and Safety at Work Regulations 1999 in the section entitled "Organising for health and safety." They refer to four key areas— control, cooperation, communication, and competency—that together promote positive health and safety outcomes within the context of the regulations. The four Cs, however, have advantages in any commercial project environment by forcing teams to recognize and validate matters concerning:

- leadership, management, and supervision;
- motivation, accountability, and sanctions;
- consultation and involvement;
- behaviour through both written and face-to-face interaction;
- capability, capacity, training, and coaching;
- selection and recruitment;
- identifying gaps in knowledge and avoiding complacency.

Consideration of the four Cs is not something that should take place in isolation at a designated point in time. In reality, we are considering these important aspects more-or-less constantly. By providing clear subject headings under which the arrangements for each project can be addressed, the four Cs method will help to define the design process landscape, providing clarity and purpose for all relevant stakeholders including the designer. Let us examine how these subjects can be appraised with reference to a design process:

Control

A well-thought-out business case will invoke control over the statement of requirements that will, in turn, control the design project as a whole, including its validation procedure. The person(s) in control of recording risk, creating the statement of requirements, directing the project, the key decision maker(s) and stakeholders, and any external parties to the project will be recorded in order to control the flow of information and the document environment.

Cooperation

The method of interaction between all parties, the points of reference, time frames, and expectations as well as how this is to be recorded and disseminated appropriately should be agreed. Internally, the organization should have a level of cooperation between the various stakeholders to ensure everyone's opinions and observations are heard and recorded.

Communication

The type, style, and frequency of documentation will have been formally described along with the location(s) of where it can be accessed, and by whom. The periodicity of meetings and the expected outcomes as well as the recording of minutes is to be formalized and disseminated so that all stakeholders know what to expect, and what is expected of them.

Competence

The abilities and proficiencies of not only those within the client's circle—administrators, managers, controllers, and so forth—but also in the wider context of the project: external advisers, for example, and suppliers or manufacturers should be reviewed against expected levels of competency. Shortfalls should be dealt with under agreed terms: for example, will additional training be carried out, or external competency be bought in. The competency of the project itself can be reviewed for capability and capacity, that is, the ability of the project to actually produce the end product in the desired quantity(ies) and within the core constraints of time, cost, and quality.

How Construction Regulations Align with the Design Process

CDM is implicitly concerned with the health and safety of persons during the pre-construction and construction stages of the project. This is through the accumulation and provision of information regarding risks to safety and the subsequent use of that information to inform the design and construction processes. Additionally, because the regulations require the provision of health and safety information in the form of a health and safety file at the end of the construction stage, they are also concerned with the operational stage of the finished product. It is because of this "forward thinking," that is, that the design should consider the product when it is in service, that CDM provides a valuable framework upon which to base the management of the design process.

Fundamentally, CDM requires the provision of the *right information at the right time to the right people* in order that they can make informed decisions regarding health and safety; whether that is from the perspective of creating the design, managing the construction site, or maintaining the finished structure. It clearly takes only a small shift in thinking to imagine incorporating all the information that should be provided across the entire project to all stakeholders, thus providing an overall management structure that has safety embedded at its core.

Benefits of Implementing CDM

The use of CDM as a model for the design process allows us a number of crucial positive outcomes, especially if the engagement is with the spirit as well as the letter of the regulations (see Table 2.12).

Table 2.12 CDM as a Template for the Design Process.

Project consideration	Benefits of using CDM as a template for the design process
Client's requirements	Although the principal designer is appointed by the client, they can be independent of the client and the designer, allowing robust interrogation of the early design concepts and requirements.
	Additionally, the provision of accurate and in-depth pre-construction information can help to guide the design requirements by considering any observable risks to the project.

(Continued)

Table 2.12 (Continued)

Project consideration	Benefits of using CDM as a template for the design process
Financial	Through the positive interrogation of the design, and of any solutions provided to mitigate risks within the project, the principal designer can advise on the correct resourcing of the project at the early stages. Increased spend on scrutiny in the early stages of design will minimize much larger spending on rectifying errors later on. The regulations also require the client to be able to devote adequate resources to the project, which can be determined from the outcome of a robust business case.
Moral	The CDM regulations are based on the framework of health and safety legislation in the UK. The provision of health and safety is an intrinsic part of societal expectations.
Ethical	The expected outcome of CDM is to create a product that is safe for all those who come into contact with it—testers, developers, manufacturers, installers, operators, and maintainers. By being required to implement the general principles of prevention, the designer will be obliged to consider the most appropriate design outputs.
Environment	Although CDM does not specifically consider the environment, the safety of the project within its location is a consideration. The assessment of salient pre-construction information will help to guide the design process in mitigating for any observable or foreseeable environmental consideration.
Regulatory	Applying the requirements of CDM to any relevant project will ensure that it is compliant with those regulations, and potentially others that we have noted.
Safety	The designer is required, under CDM, to have considered the general principles of prevention. These principles stipulate a hierarchy of mitigation techniques to reduce the risk of harm to as low as reasonably practicable. One role of the principal designer is to engage the relevant stakeholders during the design phase both with the project and with each other (the "four Cs") and this should include such individuals and groups as testers, developers, manufacturers, installers, operators, and maintainers.
Operability	The operator or final user of the product should be engaged in the feasibility stage by an astute principal designer under the auspices of the "communication and cooperation" elements of their duty. This should ensure that the operator/user is appraised early on of any fundamental health and safety requirements of the final product. Additionally, the requirement for the provision of a health and safety file, upon the product's delivery to the client at the end of the project, ensures that any residual risks—such as those concerned with operations, maintenance, cleaning, and disposal—are documented and brought to the operator's attention.

We have seen how CDM aligns with the separate requirements of regulations such as the Health and Safety at Work etc. Act 1974, as well as other standards and duties. In essence, CDM is the "bringing together" of salient parts of other health and safety legislation, relevant to the construction industry. Several pieces of legislation, for example,

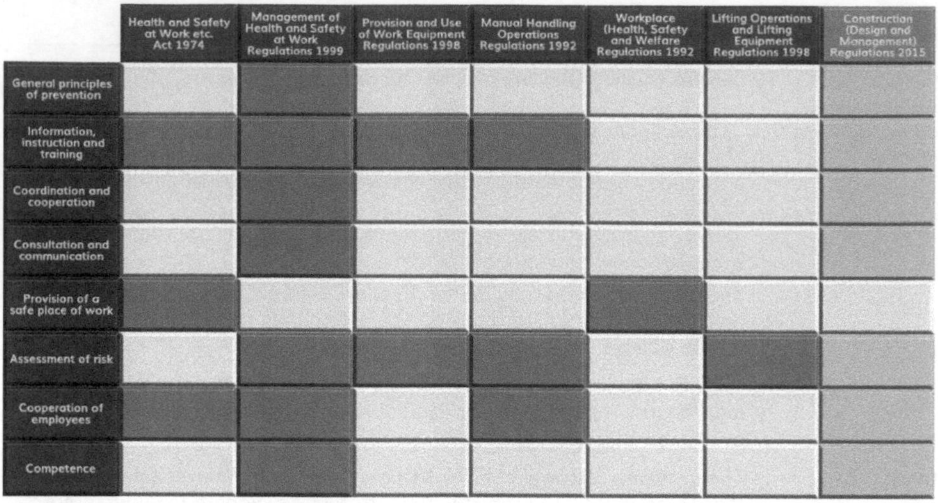

Figure 2.5   CDM Alignment with Other Regs.

cover matters like the general principles of prevention, or consultation and communication. CDM's wide-ranging requirements, comparable to the Management of Health and Safety at Work Regulations 1999, demonstrate this particular piece of legislation's remit to safety (see Figure 2.5).

It should be borne in mind that many of the core requirements of CDM are incumbent on the client, a result of their historical development over the years. The reason for making the client responsible for these requirements as a duty holder is that, as the person responsible for the financial aspect of any project, the client has the greatest level of control over how and when sufficient resources—including those connected with safety—are applied to the project in general and the design process in particular. The client can delegate the *function* of their responsibility to others but cannot delegate the *duty* itself within the regulations. The client might employ, for example, a consultant to ensure that suitable arrangements for health and safety are in place for the project but the responsibility for those arrangements (including any errors or omissions made by the consultant) remains with the client.

We have also seen how these regulations are concerned with a very much broader term of "construction" than perhaps is widely appreciated and that they are also concerned with the appropriate function of the design *and* construction (or production) process. Regrettably, there is little acceptance of the importance of the Construction (Design and Management) Regulations 2015 in many of the projects where they have relevance, particularly when one considers the large range of project types that are included by the regulations. Even within the traditional construction industry, there is often inadequate regard paid to their explicit requirements; and perhaps even less so to their implied requirements, or the spirit of them.

This is unfortunate. It is the result of political, practical, and enforcement reasons of a historical nature that are beyond the scope of this book to elucidate further. As we have seen, however, CDM provides an excellent basis for the compliance with the wider duties of health and safety legislation as well as those of other standards and duties, and the moral and ethical responsibilities of all of the relevant stakeholders in any project, not least the designer themselves. By implementing the letter and spirit of CDM, the designer will be demonstrating a level of professionalism and ethical conduct that will be reflected in the project's success, regardless of its intended output.

Essentially, CDM is about the effective management of risk through three inferred stages of a design project:

- the pre-construction information stage where information is gathered about the intended project, its application, and environment of use;
- the construction (or manufacture) stage where the product is actually made, and;
- the handover stage where the end user is provided with the product.

The regulations achieve this by imposing duties on the key participants, namely the client (who has the greatest responsibility as they control the purse strings), the designer, and the contractor(s). There is an additional duty holder—the principal designer—whose duty it is in certain construction projects to administer many of the responsibilities of the client. The principal designer is also responsible in these cases for ensuring the two other pillars of the regulations are properly carried out: that is, the effective management of the project and ensuring that the general principles of prevention are applied. This is achieved by ensuring that appropriate information from each of the above stages is passed to the right people in a timely manner and is an outcome of applying the principles of control, communication, cooperation, and competency; that is, the four Cs.

But these stages and, more importantly, this type and level of information, are not unique to construction projects. All design projects will require some form of prerequisite information; they all have a design and manufacturing stage and they all will eventually hand over the finished product to the client or end user. All projects will have certain risks that will be encountered during each of these stages. The way that information on risks is required to flow according to CDM (see Figure 2.6) demonstrates how the principle of this cascade is to reduce risks at each stage and that any risks that cannot be mitigated are passed on to the next stage until they come to rest in the health and safety file.

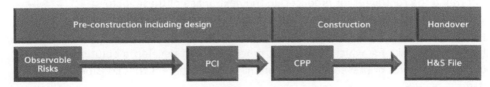

Figure 2.6 CDM Risk Capture and Management.

This is where details about residual risks are communicated to the client (or the end user) in order that they can develop their own safety strategies to deal with them. This is why it is important that all risks with the design are reduced to as low as reasonably practicable so that the end user has less risk to contend with. Regardless of the industry sector, or the type of project, this filtering and cascading method—working through the general principles of prevention each time—is an excellent basis for good design and delivery of a safe project output.

We shall now examine the phases of the Construction (Design and Management) Regulations 2015 in more detail and in particular how they can address the four Cs, which form an integral part of the management function of the regulations.

Pre-construction Including Design

The initial requirements of the regulations are to provide a solid foundation from which the design process can be managed. This requires the client to make "suitable arrangements" to provide the necessary level of resources in terms of time, money, and effort. These arrangements must then be maintained throughout the project and this can be clearly demonstrated by the use of a framework for the project whereby all the stakeholders are not only aware of their duties and responsibilities but also of the lines of communication for such things as variations, costings, design options, and so forth. A well-defined statement of requirements can also help to inform these arrangements.

The client is also responsible for providing "pre-construction information," which should identify the observable risks of the project and its intended use or location. This information is required to be provided "as soon as is reasonably practicable" and must be issued to "every designer and contractor appointed, or being considered for appointment, to the project." In traditional construction, this may include references to the ground type or composition, the method of access, and the proximity of hazards to the site such as deep water, overhead obstructions, or underground services. Where there is an existing health and safety file for a structure, the contents can help to provide information about such things as the location of hazardous materials or stored-energy systems.

In non-traditional construction projects, this pre-construction information should be considered as "pre-production information" and might include details such as the ability of the existing production facility to handle certain materials; the availability of peripheral articles or materials required by the project; and any existing hazards or risks that might be affected by the production process. Where hazards or risks cannot be dealt with through the design process, the pre-construction information will be invaluable in informing the contractor (or producer) in the next phase of these factors.

The pre-construction information should be relevant to the project in order to help either establish or further define the client's statement of requirements. It is a list of hazards and risks for which the design process should develop a solution using the general principles of prevention: it should not be a document that considers solutions itself.

During the design phase, the designer (including the principal designer if appointed) is responsible for ensuring that they take account of the general principles of prevention. As these are a requirement of general health and safety legislation and their output—health and safety information, training, residual risks, and so forth—are also requirements of CE marking, it can be seen that demonstrable adherence to these principles by the designer will have marked benefits for any project and its output.

Implementing the four Cs during this stage will ensure that the requirements of the regulations at this stage are complied with (see Table 2.13).

Table 2.13 Alignment of the Four Cs.

Requirement	Compliance—the four Cs
Formally establish the foundations of the project with regard to providing adequate resources.	Demonstrates the client's control over the project. Requires the cooperation of all the duty holders. Establishes how the duty holders will communicate with one another. Demonstrates the client's competence in providing adequate resources and that suitable competence checks have been made prior to appointing other duty holders.
Establish the statement of requirements.	Provides the control for the project to ensure the design intent is met. Requires the cooperation of the client, the client's design team, the designers themselves, and all relevant stakeholders to ensure the design output meets the requirements. All relevant stakeholders will be required to communicate early in the design to ensure their expectations are noted. By using the general principles of prevention, the designer will demonstrate their competence in providing a safe design output.
Compile and issue pre-construction information.	The pre-construction Information will allow the designer to provide adequate control measures for the risks identified. To identify as many relevant hazards and risks as possible will require the cooperation of all stakeholders and keepers of information. The pre-construction information should be communicated to relevant stakeholders in a timely manner to ensure that the design process is not delayed by matters arising. The competence of the pre-construction information should be confirmed as relevant, appropriate, and proportionate.
Apply the general principles of prevention to the project.	The general principles of prevention is a hierarchy of control to eliminate, reduce, or control risk. The cooperation of the designer and all the relevant stakeholders will be required to ensure that all risks are identified. Proposed control measures should be communicated to the relevant stakeholders to ensure there is no conflict with existing or future measures or subsequent risks. The designer must demonstrate their competence in applying the general principles of prevention in a structured, appropriate, and proportionate manner.

Construction Phase

The second phase is concerned with the safe delivery of the construction (or production) process of the product, which may include testing or development of the product due to the fact that the regulations clearly note that the design process may continue through the construction phase. The client's initial duty in this phase is to ensure that the contractor, or producer, has developed and issued a construction phase plan. In traditional construction this plan would involve such matters as site controls and rules, control measures for any hazards and risks identified in the pre-construction information, and arrangements for the safe functioning of personnel, plant, and techniques on site. In non-traditional construction projects such matters as the sequence of production; the ordering, storage, and handling of materials or articles required for production; and the site rules may be documented.

Although principally about the physical coming together of the design output, the construction phase plan (or production phase plan if you prefer) should be a formal record of how the safety of the construction phase is to be arranged and monitored as well as identifying the communication framework for design changes during this phase. Although specifically concerned with construction projects, as defined in the regulations, the construction phase plan is in essence a "safe system of work," which is a fundamental requirement of the Health and Safety at Work Act Section 2 (2) (a). Therefore, in non-traditional construction projects, a construction phase plan could potentially be seen as forming the basis of later instructional or operational documentation, especially where a product is manufactured, for example. This may assist during the last phase of the project, the handover stage.

During the construction phase, the client is responsible for ensuring that the arrangements they put into place in the pre-construction phase remain so. This will include the methods and frequency of communication between the relevant stakeholders, and the formal pathway of the decision-making process. It is important that during the production phase, any design amendments—whether required by the client or the design process itself, perhaps as a result of testing—are managed in an appropriate and timely manner to reduce the impact of delays to the project. Additionally, it may reduce or exclude the problem of other changes being required or production marching ahead on a separate element of the design while a critical design change is discussed elsewhere, which could result in costly rework being required later on.

Implementing the four Cs during this stage will ensure that the requirements of the regulations at this stage are complied with (see Table 2.14).

Table 2.14 Compliance—the Four Cs.

Requirement	Compliance—the four Cs
Prepare and issue the construction phase plan.	Demonstrates the contractor's (or producer's) control over the construction/production phase of the project. Requires the cooperation of all relevant stakeholders in the construction phase. Establishes how the design and construction stakeholders will communicate with one another during the phase. Demonstrates the contractor's/producer's competence in providing adequate control of risk throughout the construction phase.
Maintain the arrangements for the project.	Demonstrates the client's control over the project in ensuring its safe delivery. Requires the cooperation of all relevant stakeholders in delivering the safe output of the project. The construction phase plan should be communicated to all relevant stakeholders in an appropriate and timely manner. By continuing to use the general principles of prevention, the contractor and designer will demonstrate their competence in providing a safe design output.

Handover and Use

At the closure of the project, as the output is being readied for handover from the contractor/producer, the client is required to ensure that the principal designer compiles a health and safety file for the product. The purpose of this file is to provide the user of the product with relevant information about its operation, maintenance, and function. The file should also identify any residual hazards or risks that remain after the general principles of prevention have been applied during the design and construction phases. In traditional construction this file may refer to safe working loads of floors or roofs, the location of stored energy structures such as pre- or post-tensioned beams, and information about any specialist cleaning or maintenance requirements. In non-traditional construction projects the file may refer to any pertinent calculations or test results, technical reports on materials, articles, or techniques used, and instructions on operating the product during normal, maintenance, and emergency conditions.

The provision of the information in the health and safety file is entirely dependent on the project and its output in terms of its complexity, use, and the presence and severity of any residual risks. The output is similar in composition to the technical file required by CE marking and may even—if it is detailed enough—form the basis of the assessment required by the Provision and Use of Work Equipment Regulations 1998. The provision of appropriate information about the product will also align with the requirements of the Health and Safety at Work etc. Act 1974, specifically Section 2. Variously, this requires that employers ensure the health and safety of their employees "so far as is reasonably practicable" through the "provision and maintenance of plant and systems of work," the "arrangements" and "the provision of such information, instruction, training and supervision as is necessary."

Implementing the four Cs during this stage will ensure that the requirements of the regulations at this stage are complied with (see Table 2.15).

Table 2.15 Compliance—the Four Cs.

Requirement	Compliance—the four Cs
Prepare and issue the health and safety file.	Demonstrates the controls that have been applied to the product and how these may be operated and maintained over its lifetime. Requires the cooperation of the design and construction/production stakeholders to assemble the relevant information in a timely and appropriate manner. Communicates relevant operational and safety information to the end user of the product to enable them to use the product in the manner to which it was designed. Demonstrates the product's competence in fulfilling the design intent and meeting the expectations of the end user.

The Regulatory Environment—Summary

During the design process it is important to consider, amongst many other factors, not only the regulatory environment of the design process itself but also that in which the output of the process will be put to use. By considering the regulations that might apply to the product once we have completed the design will ensure that the client—as well as the end users—will be provided with a safe, compliant product to use. It is not that we are likely to deliberately make a product *unsafe* to use, but that operational, maintenance, and disposal regulations may impose certain requirements that are best resolved at the design stage. By designing-out risk at the beginning of the project we are most likely to succeed in removing risk throughout the product's life.

It can also be appreciated that of all the seemingly disparate organizations, associations, standards, and organizational objectives, the driving factors for better design are remarkably similar. They could be summarized as:

- reducing errors;
- reducing costs, and;
- reducing waste.

Clearly, the main reason for increased cost and wastage is the occurrence of errors; be they design errors, communication errors, or production errors. By utilizing the requirements of CDM—both in the letter and spirit of the law—we have seen that a strong, solid foundation for any design project can be gained. The general principles of prevention provide us with a fitting template for reducing risk, not just at the design stage, but throughout the product's life, when applied correctly. And the four Cs of control, cooperation, communication, and competence help to ensure that all the right information is passed to the right people at the right time during the project.

The reduction of errors during the design stage will inevitably reduce costs, loss, harm, and wastage for the entire duration of the design output. In doing so, it will be useful to consider the following:

1. The regulatory environment in which the product will be used to ensure the design creates the safest "whole" solution.
2. Whether maintenance or repair processes subject to separate regulatory controls?
3. What the risk tolerance levels are of the marketplace into which the product is to be placed?
4. Understand the client's reputational risk tolerance:[1] do they expect to abide by relevant regulations or wish to exceed the regulatory requirements?
5. Thoroughly documenting the design process to allow the creation of the most complete technical file possible for conformity purposes.
6. Ensuring that signage, wording, and instructions comply with relevant regulations as well as being clear, concise, and appropriate for the end user or marketplace.
7. Ensuring that suppliers and/or producers engaged in the project are compliant with regulations; that is to say, demonstrate their competence in terms of materials, management, procedures, techniques, and so forth.
8. Developing a thorough pre-construction document that examines the core constraints of the project and establishes the environment of the design process (the four Cs).

---

1 Risk tolerance is a theme we shall examine in detail in the risk management chapter.

3

Design Process Considerations

Management Structure and Delegations

As part of the management system, it is important to include a definition of the management structure not only for the design process but also for the project itself. This structure will help to establish the decision-making process as well as the responsibilities of each role identified in the project. It should also reflect the line management roles so that individuals are aware of to whom they report. Being cognizant of the line management of a client's or contractor's organization may not seem immediately beneficial to a designer, but understanding these levels of authority can be crucial. As a design matures, and more stakeholders are potentially involved in the process, there may be requests for alterations or specifications received by the designer from individuals or groups, which may impact, perhaps greatly, on the project's time, cost, or final intent. Being able to identify the level of authority from whence such requests came can at least allow the designer to confirm their validity.

With any project, as with any organization, there will always be a key decision maker, and their role is vital in signing-off critical elements of the design as well as the final design prior to the production stage. In a relatively modest project this may be straightforward as this key decision maker will most likely be involved to a high degree in the design process and its outcomes. In large or complex projects, however, the key decision maker may be removed from the day-to-day function of the design process, possibly by several layers of intermediate management. Here, therefore, it is vital that the decision-making process is accurately documented and controlled. The key decision maker will have a responsibility to their own stakeholders—perhaps the organization's board, or its shareholders—and is therefore owed an obligation to have the most up-to-date, accurate and complete information made available to them.

The management structure may well include signing authorities or clearly defined areas of responsibility. Signing authorities are often used by very large organizations to determine the value that individuals can make procurement decisions up to and are sometimes established within their job title's description. Areas of responsibility differ slightly in that these are usually established around particular functions or areas of the project. The key decision maker may, for example, be concerned with the absolute cost

An Effective Strategy for Safe Design in Engineering and Construction, First Edition.
David England & Dr Andy Painting.
© 2022 John Wiley & Sons Ltd. Published 2022 by John Wiley & Sons Ltd.

of the project as well as ensuring the design intent is fulfilled; they may devolve the responsibility of what the door handles are made of to someone else.

Within the design organization, if there is one, the line management structure should already be established for the effective management of the human resource. The structure for the project may not, at this point, be so well established, if at all. It is, however, important to address and document the fundamental roles within the project early on so as to establish the channels of communication from the outset. Later, various sub-roles and duties can be added as required, depending on project complexity.

From the designer's perspective, understanding the client's management structure, and juxtaposing their own with it, is relevant to ensuring the design process does not stall when decisions need to be made or when emergent issues need to be addressed. Not every design or procurement decision requires the convening of a meeting to discuss it. But even if the designer is unsure who makes particular decisions, they should have a clear indication about to whom an approach should be made in order that the correct decision maker is engaged. This may be through the project manager, programme manager, procurement manager, or other role holder.

The size and complexity of the design management structure is wholly dependent on the size of the design organization and the type of work that they are undertaking. At its very basic level there should be a single point of contact at the top of the structure. Ideally this person should sit on the executive board, or at the very least should be part of the senior management team. Without this lever it may be difficult to make critical decisions without having to ask others. If it is a relatively large company, this person should delegate their authority to suitable candidates. With this delegation of authority should also come the ability to make independent financial decisions, possibly up to an agreed level.

Again, the size of the business may require several layers of design management. Each of these layers should be delegated an agreed and managed level of responsibility. For a smaller business it is perfectly acceptable to undertake a dual role with multiple accountabilities as long as that person understands the design process and is able to make relevant decisions.

Client Relationship

At the beginning of any design process there will be an originator who may be an individual or a corporate body or possibly another design process altogether that drives an additional or supplementary requirement. For example, during the space programme begun by the National Aeronautical Space Administration (NASA) in 1961, there were many supplementary design requirements peripheral to the main ambition of placing an astronaut on the moon. The originator of any design requirement is the client and, whoever or whatever, represents that client, their requirements will be framed by the same core factors in each case, which are:

- time constraints;
- financial constraints;
- quality and operability considerations.

These core factors will vary greatly project to project, and the designer must be fully understanding of their relevance and importance to the client. It is entirely possible that any one of these three factors will be of more importance to the client than the others and this, too, is of relevance to the project as they will undoubtedly have an effect on the design process. Constraints on time may require the design to be completed by an increased number of designers, possibly causing errors or confusion to occur. Financial constraints may lead to savings being made that affect the quality, safety, or operability of the final product. And quality demands may require the introduction of materials or processes for which the designer had not accounted, or has little relevant experience.

Where there is a single individual as the client, it is evident that the designer should have a straightforward relationship during the design process. Where the client is a corporate body, or other group of individuals, it can be surprising, however, how the design requirements for an even relatively simple product can quickly become beset with elaborate or even conflicting expectations. This can be particularly evident where the client, as a corporate body, has little or no internal control of its executive function, or where disparate departments are uncommunicative or even truculent with one another. It is not the position of the designer to ameliorate their client's internal dynamics, nor proffer advice on its level of functionality (or dysfunctionality!) but it would certainly benefit them to at least have an understanding of their client's decision-making process.

Where the client appoints a project manager, or perhaps a principal designer, this can often assist with providing a single contact point through which to channel the flow of information back and forth. It can also be valuable to understand the motivation behind the client's requirements: this may be driven by a personal, departmental, corporate, or perhaps more ethereal need.

In traditional construction there are two principal ways that projects are often contractually arranged: "design *and* build" and "design *then* build." The differences have no impact on the methodology described in this book, but it will be of interest to those readers looking to engage in this sector.

Design and build is used to describe those contracts where the main contractor for the work is to take the design and turn it into a working product themselves. This may mean that the design is only at the concept stage at this point, or it may mean that the specifications have been defined but the technical aspects of the construction are to be determined by the contractor. Alternatively, the contractor may be engaged from the very beginning of the project and employ their own design team (or a subcontracted one) for the entire design process. However the contract is arranged, during the construction phase the contractor assumes all liability for the design and any amendments arising from it.

Design then build follows the more traditional path of a designer being engaged to complete the entire design process and hand over a completed full design to the contractor for construction. During the construction phase, the designer assumes all liability for the design and any amendments arising from it.

Whichever way a construction project is completed, the best practice methodology for any construction project should ensure the following are given appropriate consideration:

- the management structure, the documentation environment, and the engagement of stakeholders;
- the implementation of the general principles of prevention;
- the identification, recording, and management of risk;
- the timely provision of salient pre-construction information;
- the engagement of relevant stakeholders.

As well as the factors driving the client's requirements, the client's structure may also have a bearing on the design process and it can be useful to understand this, again through asking salient questions at the beginning of the project. Confliction among corporate bodies or groups of individuals is not uncommon, but it is the process of dealing with that conflict that is important. In any relatively well-established organization, there may be many departments who exert an influence on the project for a variety of reasons (see Table 3.1).

Table 3.1 Influences on the Project.

Department	Possible/hypothetical reasons for having an influence on design
Board of directors	The board will be aware of the long-term ambitions of the organization; perhaps its sale or merger with another company. It has a responsibility to the shareholders to secure profitable returns on any investment.
Finance	This department is concerned with levels of indebtedness and has a responsibility to the financial director to secure the best value for money. They may be aware of financial pressures that mean the organization must minimize investment wherever possible.
Procurement	This department has a responsibility to source the most cost-efficient solution to any investment and may have established a quid pro quo with a supplier that will affect the sourcing of certain materials/equipment.
Sales	The sales team will have detailed records of the organization's customers and their tastes and demands. They may know of customers who are moving in a particular market direction away from the organization's current stance, and they will have a responsibility to deliver a product that meets existing and future demand.
Marketing	Marketeers will know what features of previous products were market-leading or well received by customers. They may also know what features will be demanded in the future and how the organization's product is compared in the market to others in terms of quality, function, and price.
Manufacturing	The manufacturing department will have records of how previous products have been manufactured in terms of cost, time, and quality. Combined with this will be detailed knowledge of how existing manufacturing equipment, techniques, and personnel can achieve any given product design. They have a responsibility to deliver the best cost price per item of each finished product.

(*Continued*)

Table 3.1 (Continued)

Department	Possible/hypothetical reasons for having an influence on design
Quality assurance	The QA department will be aware of previous product failures and issues as well as knowledge of what ex-customers demanded from previous products, which if introduced in a new product would return the customer to the organization.
Operations	Operations will know how previous products could be maintained and interacted with by the organization and the customers. They have a responsibility to improve cost savings across the organization and will have an awareness of any new machinery/equipment that is available or is to be procured by the organization in the near future.
Maintenance or technical support	The maintenance team will understand the issues with maintaining and repairing previous products or with the equipment that manufactured/installed/modified them. They will have details of the skills level of the personnel who will be performing these tasks on future products.
Shipping	The shipping department will be aware of any issues with transporting previous products and any factors that need to be taken into account. They will also be aware of any relevant existing and future regulatory control over the import or export of materials or equipment.
Warehousing	The warehousing department will have details of the quantities and format that best suits how products are stored and transhipped.
Human resources	HR will have records of the duration and reason for any absences connected with interacting with previous products.
Health and safety	Health and safety will know of any apparent safety risks with previous products and what mitigation or control measures were required. They will know what safe working practices are required and how, if at all, personnel have developed stratagems to overcome any inherent risks.

This is not to suggest that the designer engaged on any given project should, or would even be able to, avail themselves of all the possible machinations that have led their client to the point of requesting a designed output. But this list gives some indication of the myriad possible causes for, firstly, the design requirement and, secondly, the possible reasons for changes to the design requested by the client during the design process. Understanding these possible economic, strategic, financial, and social input factors can be pivotal to creating the basic framework around which the project can progress. Anticipating the potential risks to the project is vital to developing a robust risk register and design risk assessment, and the examples listed above are but one element of those risks.

One example of the complexity of the client's relationship, not only with the designer but also within their own internal organization or group of stakeholders as well, is that of building a warship. The client, in the UK at least, is The Crown, an indistinct body that represents the state and its national interest. The Crown is represented in person by the Monarch and is administered by Her Majesty's Government, which is composed of elected members of parliament who are assisted by civil servants employed by various departments of the government. One of these being the

Ministry of Defence, whose job is to procure for The Crown suitable equipment required for the defence of the realm. This procurement is overseen, however, by the Treasury, whose function is to keep an appropriate hold on the nation's purse strings. Both the Treasury and the Ministry of Defence are overseen by the National Audit Office, who are civil servants tasked with ensuring The Crown gets value for money, and by Parliamentary Select Committees, composed of MPs, who want to ensure their own re-election!

And, although the Ministry of Defence is the client's lead in the design, the actual output—a warship—will be owned (technically) and operated by the Royal Navy who have very much their own command structure, including their senior management known as The Admiralty. On the warship itself there may be not only Royal Navy crew but also members of the Fleet Air Arm (who are part of the navy), the Royal Marines (who are an armed service in their own right), or the Royal Air Force, who will all have their own systems, procedures, and requirements. It may be seen from this example why the procurement of military equipment can be such a long, complex and, above all, expensive process.

The relative importance (the relevance) of all these factors connected with the design should be noted early on during the design process and can rarely be done other than through astute questioning of the client's needs. Establishing these factors, along with the reasonings behind them—in terms of the core constraints of cost, time, and quality—will ensure the design process has a well-defined landscape in which to flourish. Equally, where deviations become apparent, they can be measured against the initial expectations and, where a client suggests that the design process is not as they originally expected, or where they introduce a change in the requirements, these can be quickly reconciled and, more importantly, costed appropriately. It should be remembered that anything is possible, but always at a price.

The designer cannot sit as referee to the internal prevarications of their client but must instead produce the output in accordance with the contract to which they have been appointed. However, awareness of the environment from which the intended design stems should help the designer produce a better output and, additionally, possibly present solutions at the concept stage, which may open the discussion up with the client about what their actual requirements truly are. This will have a positive effect on the designer's reputation, professionalism, and competency.

It is not unusual for a design process to alter over time, especially on long-term or complex projects. Variances in requirements may become necessary for any number of reasons we have already discussed, but these should always be considered against the original core constraints of cost, quality, and time. The client is free, of course, to amend the design requirements as often as they choose but by "freezing" the design, during the latter stages of the design process, the client can be made aware of the impact on those core constraints. By having a formally-defined, structured landscape of the design process with clear expectations, both client and designer will be protected from any negative impacts that variations can cause them.

As the design matures, through technical studies and on to the final detailed output, it is important that the relationship with the client is maintained and that the questioning process continues. Often in large or complex projects, there will be an appointed person to provide this continuity between the client and the designer or design team. Under the Construction (Design and Management) Regulations 2015 (CDM) there is a requirement, under certain project types, for there to be a principal designer. Despite the title, this role is not primarily concerned with the actual design of the project but with the involvement of all of the stakeholders with a view to ensuring their continued engagement and the subsequent flow of information.

Documentation and Management Systems

Regardless of its size or complexity, an important component of any design project is the method of controlling and documenting the information and communication that it generates. Large organizations often already have in place a standardized system to which their documentation and management systems adhere. On large and demanding projects there may well be in place an individual, or even a small team, whose sole task is to monitor those systems and disseminate information in an appropriate manner. Systems may be based on such standards as BS EN ISO 9001:2015 Quality Management Systems (ISO, 2015), MSS 1000:2014 Management System Standard (CQI, 2016), or PAS 99:2012 Specification of common management system requirements as a framework for integration (BSI, 2012). The management of a design project, from a design perspective, is not the same as project management as conducted by the project or programme manager. The designer — or design team — must administer the flow of information back and forth between client, stakeholders, and designers in a way that ensures only the latest information, acknowledgements, decisions, and objections are being worked on. Where a management system provides the *strategy* for managing the information of the project, the documentation management system addresses the *technical specificity* and *chronology* of that information.

In this digital age, most organizations rely on cloud- or server-based technologies to distribute electronic documentation. It can, however, also cause confliction errors where a document is being worked on or accessed simultaneously by two or more users. The problem of confliction has become widespread since digital integration and requires an absolute solution. Some proprietary software systems enable users to "lock" documents while they are editing them, or they may provide users with the ability to view and download documents but not to edit or upload them.

Whatever system of document management is used, it must, like the overall management system itself, be robust enough to deal with not only day-to-day issues and misunderstandings but also exceptional circumstances, such as power failure, loss of data, infringement of copyright, and unauthorized access and editing. Again, large organizations may have in place a role assigned to an individual for disseminating and controlling documentation that may form part of the management system standard that the organization uses. The system should also ensure that the correct stakeholders not only have access to the documentation but also have the correct levels of access. This can often be agreed at the outset of any project during the definition of the management structure. It is also important that the designer or design team has/have an area within the documentation system where ideas, concepts, and sketches can be freely discussed by the designers alone. Sometimes referred to as a "sandbox," this area should not be available to any other stakeholders due to issues with misunderstandings, confidentiality, impartiality, or plagiarism.

With the prevalence of technology comes another issue for the reporting of meetings. When individuals meet physically for a meeting there is often one person assigned the duty of taking minutes that can be disseminated later to the attendees. With online meetings this can be easily overlooked. Some online meeting software allows the recording of meetings that may, of course, allow the transcription of minutes at a later stage. The recording of individuals does, however, bring about its own issue of privacy and data protection. If one of the attendees objects to the recording of the meeting then they may withdraw from the consultative process and their potentially valuable opinions will be lost. Consideration of the method of taking minutes of meetings is extremely important and is not to be left until the meeting has started or, worse still, ignored completely.

Building information modelling (BIM) provides guidance on how a project should organize and digitize information, down to the metadata a project should use to allow for easy identification of the document owner and relevance to each stage of a particular project. This is a useful system to adopt particularly for larger or more complex projects where there can be many drawings and a large amount of information.

Communication and Dissemination

Within the management system of the project it will be important to establish who all the relevant stakeholders are and, if possible, their own objectives and interests in it. This differs from the structure of responsibilities in that many of the stakeholders may not have a direct part to play in the development of a project—making decisions on the finite details for example—but their input may be equally valuable, if not crucial. It is important that the various and sometimes disparate stakeholders are involved in the design process as they may bring significant details to the project in terms of their own influences. Often even large-scale projects can be concluded to the satisfaction of the client, the designers, and the key stakeholders, only to find that the people who are to engage with the output every day are frustrated by simple elements of the design that could have been readily modified had they been known about during the design process.

Within the designer's own organization, it will be readily straightforward to identify the various design disciplines that will engage with the project—mechanical, electrical, architectural, environmental, civil, and so forth—but, as with the management structure, it may be less so with the client's stakeholders. There may even be reluctance on the part of the client to identify the individuals or groups that make up the entire relevant stakeholder list. This is, of course, the prerogative of the client. However, it is highly valuable, and indeed desirable, for the designer to understand the underlying intention of the design. Consider a new berth being built at a port where the key decision maker is basing their requirement for berth capacity on a current list of vessels, whereas the operations manager is working towards accepting much larger vessels with a view to increasing port throughput.

These types of issues should be identified early in the project of course, but if the operations manager in this case is excluded from early meetings, it is entirely feasible that the design process could get to a significant stage of maturity before this additional requirement is revealed. Hence, the designer should be searching for these issues from the outset in order to provide the client with the best possible output with the least amount of wasted effort in variations and, of course, errors. There is also the intrinsic safety of the project's output to consider too. It is only by assessing the needs of the shop floor operatives, via their line managers or departmental heads, that the design process can truly consider any risks that may be readily eliminated in the final output. By eliminating — or at the very least, mitigating — these risks, the output will greatly benefit all those who have to interact with it.

In identifying the relevant stakeholders in the project and their respective intentions and needs from it, it may be appreciated that this will also enable the *right* information to be passed to the *right* individuals at the *right* time. We discussed earlier how the key decision maker may delegate responsibility for a particular element of the design (in this case, we mentioned earlier about door handles). With this knowledge, the designer can ensure that communications regarding either the project as a whole, or just about the door handles in particular, are directed to the appropriate individual promptly, thus saving time and reducing wasted effort. We can also see that, by having an understanding of the intentions of various stakeholders, as well as their signing authorities, the designer can quickly establish the appropriate action if a requirement is made for door handles of an exotic material at an inordinate cost. By also having in place a well-defined document management system, the designer is easily able to demonstrate their reason for querying this request, as well as the route that was followed in doing so (in terms of line management), and the response that was received and any action that followed.

At the very start of every design project, it is important to understand who all the stakeholders are, what they do, what they are to provide, and what they need. This, as well as the periodicity of information flows and specific design management meetings, will assist in the scoping and definition of a communications plan. Certain project management methodologies identify this step at the very start and ensure that all communications are managed appropriately.

It is up to the stakeholder group to decide how often they want to meet, what can be agreed in which meetings, and who has to be present, although it is also important to agree who is required to make any meeting quorate. Meetings held without one or more of the key decision makers present may not only be a waste of valuable time but may also arrive at decisions that are later countermanded. The risk is also that if design drawings are completed or, worse still, if production commences, there might be a need to redevelop those drawings or rework the product, costing money as well as time. Deciding, therefore, who makes any design meeting quorate or not is not simply polite but actually critical to the safe and effective delivery of the project.

It is beneficial to develop a communications plan that covers all of this and also useful to develop terms of reference for all stakeholder meetings. This ensures that a meeting held initially to decide, say, the colour of the walls doesn't end up deciding to add another two storeys to the building. A typical stakeholder communications plan will identify all the meeting types, their periodicity, and who should attend. It is good practice to include any preparations stakeholders need to undertake prior to the meeting and how the output of the meeting will be disseminated.

Generally speaking, a communications plan should include the following:

- A list of the meeting types to be covered by the plan: for example, project design update; project planning meeting; daily briefing; lessons learned; and so forth.
- The format of each meeting, whether online, face to face, via telephone call, and so forth.
- The anticipated frequency of communication, that is, weekly, fortnightly, monthly, and so forth.
- The name of the person chairing the meeting. Alternatively, a job title could be cited to allow for delegation in the event of absence.
- The stakeholders who must attend at each type of meeting. This ensures that each meeting is quorate.
- Details of any preparation required for each meeting, that is, any documents that need to be read or commented on, or any that need to be submitted for discussions prior to the meeting, and when this has to be done by.
- Whether the meeting is minuted (or recorded) and who is responsible for that.
- What the desired output of the meeting is: for example, a summary report to the board, the production of metrics, and so forth.

Project Management Methodologies

The association for project management (APM) defines project management as the:

> "Application of processes, methods, skills, knowledge and experience to achieve specific project objectives according to the project acceptance criteria within agreed parameters. Project management has final deliverables that are constrained to a finite timescale and budget.
>
> A key factor that distinguishes project management from just 'management' is that it has this final deliverable and a finite timespan, unlike management which is an ongoing process. Because of this a project professional needs a wide range of skills; often technical skills, and certainly people management skills and good business awareness."
>
> *(APM, 2019)*

There are a number of project management methodologies, designed for use with different types of projects. We have chosen to name only a few so as to show you the variation and what can be covered; these include the following:

- RIBA plan of work – design and process management tool for the construction industry.
- PRINCE2 (Projects IN Controlled Environments) – this is intended for large and complex projects and has very specific requirements for the management of the project.
- Agile – designed specifically for software development projects.
- ISO 21500:2012 – generic guidance on best practice in all types of project management.

- Association of Project Managers' Body of Knowledge – foundation guidance for project managers in all sectors and industries.

The role of project management, per se, is not the intention of this book. As such, we have chosen just two of the above examples to discuss in general terms.

RIBA Plan of Work

The Royal Institute of British Architects (RIBA) have their own methodology for managing the design stage, which, whilst reflecting the ambitions of CDM, provides some additional, and more defined, stages to the process. And while it may be assumed, due to the inclusion of the word "architect," that this methodology is concerned with "traditional" construction, it may be seen from the explanation below that, in fact, this sequence could readily be applied to virtually any design process. We can see how these more defined phases of the Plan of Work correlate with the broader definitions of CDM in comparison to the design process (see Figure 3.1).

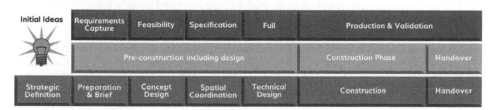

Figure 3.1 Design Process (Top), CDM 2015 (Middle), and RIBA Plan of Work 2020 (Bottom).

The RIBA Plan of Work 2020 (RIBA) is concerned with the management of the design process rather than the management of the project itself and consists of eight stages during the design process (see Table 3.2).

Table 3.2 RIBA Plan of Work 2020 Stages.

RIBA Stage	Details
Stage 0 – Strategic definition	This is the very early thinking about what the client needs the design to provide. This stage is normally conducted internally without the need to employ a designer just yet.
Stage 1 – Preparation and brief	This stage aligns with the design life cycle requirements capture stage; it checks that the idea of the design is viable and helps to provide the direction that will be needed in the next stage.
Stage 2 – Concept design	This stage aligns with the design life cycle concept feasibility stage and is the last time to check that the design concept aligns with the client's vision and budgetary expectations.

(Continued)

Table 3.2 (Continued)

RIBA Stage	Details
Stage 3 – Spatial coordination	This stage aligns with the beginning of the design life cycle specification stage and involves putting details into the design and undertaking the relevant calculations to ensure that the proposed design solution is possible, safe, and within budget still.
Stage 4 – Technical design	This stage aligns with the final stage of the design life cycle and involves the provision of information with enough technical detail to construct the whole design.
Stage 5 – Construction	This stage aligns with the design life cycle production and validation stage and is mainly about managing the design as the construction progresses. With complex designs, this stage may involve lots of incremental changes to the original design as issues are identified and the design is modified.
Stage 6 – Handover and close	This stage aligns with the design life cycle design acceptance stage and is the result of completion of the agreed validation checks/tests. With complex designs this may run concurrently throughout the construction stage, where the documented acceptance information is used to develop safety justification for the continued construction of the design.
Stage 7 – In use	This stage aligns with the design life cycle in-service stage and is associated with the operation use of the design. The project normally stops at the end of RIBA Stage 6, although there may be warranty issues for both designers and contractors involved in the design/build. If there is a requirement to implement mid-life updates to the design, this may involve the original designers and construction contractors in an advisory or perfunctory capacity.

PRINCE2

PRINCE2 stands for **P**rojects **IN C**ontrolled **E**nvironments and was designed to bring a structured, uniform pathway to project management—originally for very large government contracts. The project manager is required to have completed a PRINCE2 Practitioner course and other key stakeholders must have completed a PRINCE2 foundation course in order to have a good understanding of the processes involved and what their accountabilities are. PRINCE2 can be applied to any type of project within any project environment (see Figure 3.2).

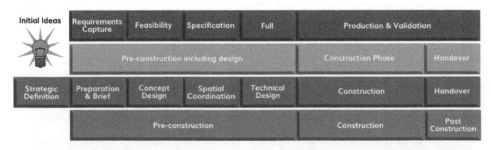

Figure 3.2 Design Process (Top), CDM 2015 (Upper Middle), RIBA (Lower Middle), PRINCE2 (Bottom).

   This methodology provides clear instruction for each stage of the project and, if followed properly, will result in a well-managed project. It is quite different from the RIBA Plan of Work as it is not specifically there to manage the design; rather, it is there to make sure that the project is managed correctly from a time–cost–quality point of view and any issues are captured and used appropriately to inform executive decisions. To this end, it can be seen that the requirements of information gathering, recording, dissemination, and control—all key to good, safe design—can be readily managed by this methodology, in any size of design project.

Environmental Impact and the Circular Economy

It is not the intention of this book to detail environmental design or specify the tools that are available to assist with Ecodesign. It is sufficient that we discuss some of the attributes that virtually every design must take into account in terms of environmental impact. Indeed, there is much environmental regulation now, across a wide range of markets and territories, that must be factored into the influences that affect the design process. These should not be considered as the aim point, however, but as the baseline for our designs and production processes. We have examined the various stages of the product's life, and how design is of paramount importance in defining every stage, in terms of operation, maintenance, and disposal. Moreover, the design of any product can affect the environmental impact of the product as well *during these stages*. From the energy that the product requires to function to the types of chemicals that clean it and lubricate it, to the effort required to dispose of it, and the substances that we are left with afterwards.

   Environmental concerns are not "somebody else's problem." Every human on Earth is a stakeholder in the concern for the environment because every human calls this planet "home"; and at current rates we are far more likely to destroy it than find a way to inhabit a new planet and destroy that too. The planet may be paying the price for Western profligacy but it has become everyone's problem to try and resolve. This is why every stakeholder in a project should equate environmental concern with the product as highly as they do personal safety and financial judiciousness. Some of things to consider are listed below.

- The environmental impact at the very earliest stages of the design process.
- Identifying the appropriate legislation/regulations/standards and adopting them as a baseline; that is, having the expectation to exceed them wherever possible and practical.
- The supply chain in terms of:
  - the extraction and processing of raw materials;
  - manufacturing techniques and location;
  - the transportation or logistics of moving materials and products efficiently.
- The use to which the product will be put in terms of energy use and sustainability.
- The ultimate disposal of the product including how this is likely to be done and where; and what will remain afterwards in terms of waste.

These considerations tie-in with the five main phases of the environmental life cycle of any product and the energy and materials required to make it: extraction, manufacturing, transportation, use, and disposal (see Table 3.3). It may be seen that these considerations align neatly with the product life cycle for which we are designing and thus we can readily append them when initiating the design process:

Table 3.3 Environmental Life Cycle Phases.

Phase	Design considerations	Environmental considerations
Initial ideas	What needs to be designed and why?	How can it be sourced more sustainably, ethically, or responsibly?
Design	What is the market, the purpose, and the function of the design?	How can it be brought to market in a responsible way?
Production	What is the capability and capacity of the manufacturing set-up?	Are there alternative materials or manufacturing processes available that reduce the environmental impact?
Use	How will the product be used, maintained, and repaired?	Can we make its use more sustainable? Can we make any consumables less impactive?
Disposal	How is the product likely to be disposed of and will this expose the user to harm?	Can we reduce the amount or type of waste the disposal process creates to prevent environmental harm?

Factoring in environmental concerns is not to be seen as a "bolt-on" to the core design process: it is incumbent on the designer in particular to ensure that these factors are pushed to the very forefront of the initial discussions with the client. The designer should be in a position to promote sustainable and ethical solutions of which the client may not have been aware, obviously within the scope of the statement of requirements. And, as with all of the many considerations that the design process must take into account, solutions to environmental factors are most cost-effectively dealt with in the earliest stages of the design. Confronting and resolving issues early will always save redesign, rework—and expense—later on.

As with risk, where introducing one control measure to prevent a particular risk can invoke a new risk becoming apparent, attempting to resolve an environmental concern can often involve some form of trade-off. The global push to produce electric vehicles, for example, has resulted in an exponential rise in the mining of the metal lithium for their batteries. Apart from the physical impacts of extraction, such as the energy consumed and devastation caused, the logistics of moving large amounts of this element are also of concern, especially as the great majority of working mines are located in Australia, South America, and China. Clearly it would be churlish to ignore the positive impact of using a rechargeable battery to propel our design in favour of some fossil-fuelled alternative, or a wooden-framed building over steel and concrete, but it can be beneficial to consider both the material and supply source.

The supply chain is, like many other sectors, devoting ever more effort to reducing its environmental impact. There are trade and material standards that signify a supplier's adoption of certain environmental criteria, such as the Forest Stewardship Council's FSC certification of wood, for example. Investigating any potential supplier in terms of not only their competency, capability, and capacity (to supply the right material or component at the right time in the right quantities) but also the mitigation of their environmental impact, is crucial in the additional stakeholder phase of the project. Depending on the complexity of the project of course, this may also involve the procurement team: hence why it is important to have acceptance by these and all other stakeholders of their environmental responsibilities.

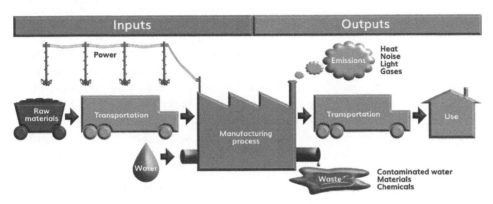

Figure 3.3 Environmental Inputs and Outputs.

There are several environmental factors associated with the inputs and outputs of the design process (see Figure 3.3). This is but one simple iteration and would of course be multiplied for more complex projects. A factory producing plastic products would, for instance, receive raw material—in this case plastic granules—at one end of the factory and produce finished, palletized products at the other end. In the case of our home printer, however, there may be many components involved in its assembly, possibly made by a variety of manufacturers, and these may then be assembled elsewhere. This is certainly the case with many warships: the Royal Navy's Queen Elizabeth-class aircraft carriers were produced in six separate shipyards, which themselves contained many manufacturing and assembly areas. The inputs shown in the diagram include raw materials (and transportation) and additional resources required for manufacture such as water (cooling water, wash down, etc.), oils, greases, and power (electrical, gas, oil, diesel, etc.). The outputs include the transportation to point of sale and all forms of pollution that the manufacturing process generates, including waste streams (water possibly contaminated), heat into the atmosphere, or water used for cooling, light pollution, and noise pollution.

In "traditional" construction projects, that is, buildings, there are additional environmental concerns that will need to be considered, such as local flora and fauna, and possibly sites of special scientific interest (SSSI), as well as the direct impacts of the construction process. It is important to also consider the use and delivery of raw materials, possibly using local products to reduce the distance travelled. Some construction is now achieved through the use of prefabricated segments of the structure that are joined together on site. Whilst this can achieve a high level of consistency in the construction, as well as potentially saving time and costs, it will invoke other considerations such as transport to the site and craning the segments into place.

For traditional construction projects too, there will be a requirement to conduct an Environmental Impact Assessment (EIA), which is designed to inform planning authorities of all the significant effects that a proposed project may have, and the mitigations that will be put in place to ensure that the risk to the environment remains within acceptable limits. The EIA is also used to inform the general public, so that they may meaningfully participate in the decision-making process. The Town and Country Planning (Environmental Impact Assessment) Regulations 2017 (2020) defines the types of development that require an EIA.

The Circular Economy

One hundred years ago, things were made to last. Not in the sense that they were made better than they are today (although that is a moot point), but in the sense that many products were in fact designed to be reused many times over. Milk was delivered to houses in glass bottles that were endlessly recyclable. Brake shoes and clutch plates on cars were designed to be relined with friction material. Clothes would be made with extra material in the seams so they could be sent to a tailor or seamstress to be adjusted as the owner's body shape changed either through nourishment or age. Children's clothes would be handed down to younger siblings as the original occupier grew out of them. Ultimately, clothes—along with all manner of other disposed-of items—would be collected by rag and bone men who would convert their use or their ownership. What we would call recycling and "upcycling" today. In fact, rag and bone men, together with that most environmentally-friendly mode of transport of theirs, the horse and cart, were still to be seen on the streets of London as recently as the 1990s.

The phrase "make do and mend" was used by the UK government during the Second World War to highlight the need to reduce waste, and the saying "waste not, want not" (i.e., using resources carefully so as to prevent being in need of them in the future) can be traced back to the late 1700s and was quoted in a novel by Charles Dickens. The reason for all this probity may have had more to do with saving money than environmental altruism but it still helped to reduce waste. The Industrial Revolution may well have been the catalyst for the exponential demand for fossil fuel, as well as ushering in a time where humans were no longer the primary source of production, but it was the introduction of mass-production techniques in the United States, principally by Henry Ford, that led to burgeoning American consumerism in the 1950s that drove us to a world today of throw away products and greed over conscience. Today we are, thankfully, far more conscious about the waste we produce and the energy we consume. This has come about by a mixture of public awareness, governmental action, and corporate responsibility. It has been assisted by the integration of disparate peoples across the globe, once separated by distance, now freely able to communicate in real time using the internet.

The modern economic model is a linear one, where raw materials are converted into products that are used for whatever purpose and then discarded. This is true even of pre-consumerism times where products were reused or repurposed many times before being finally thrown away. The natural world, however, relies on a cyclical system where plants are eaten by animals, who in turn are eaten by other animals who ultimately die.

All the waste produced or excreted by the whole process becomes soil and nutrients, which, in turn, allows new plants to grow. Just as in physics, where we know that energy is not consumed but simply converted into another form of energy, the natural world constantly recycles material in a circular fashion.

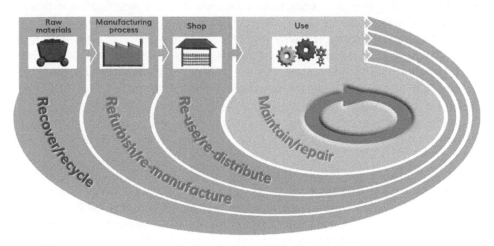

Figure 3.4   Circular Economy.

The current thinking around the "circular economy" (see Figure 3.4) is, ostensibly, a return to more practical times before ostentatious consumerism. It is driven by the need for all advanced economies of the world to do *something* about the vast amounts of waste they produce rather than saving a few pounds on having to buy a new pair of shoes. In fact, the principles of the circular economy may well cost more in the short term as systems and conventions are brought to bear across the global economy. This cost, however, can be set against the cost of not having a habitable planet on which to live.

The quintessence of the circular model is that, rather than buying, say, a washing machine and throwing it away after a few years and buying a new one, it would be completely recycled or refurbished for someone else to use. The new machine would itself be made of recycled components and materials. Further still, the products you buy could instead be rented from the manufacturer to whom it is returned after a period of use; again, to be recycled or refurbished. Packaging too, can be minimized, returnable, or made from recyclable or even compostable materials. Small but significant steps are being made in this direction all the time. The motor car industry has, for example, been making great strides in using ever more recycled material in the production stage and advanced recycling centres have been set up to extract ever greater amounts of material from scrapped vehicles.

The part that design plays, for the reasons we have stated, is key to providing design solutions that return to the considerations of durability and repeatability. That is to say, how can the product be used and then reused or repurposed as efficiently and as elegantly as possible. This is, of course, a deviation from the original design intent: however, if the route to and through disposal of the product is *designed in* at the earliest stages of the design process, then this deviation *becomes the design intent*. It is then a matter of maintaining this intent throughout the product's life cycle. For example, a product that acts as a delivery device for something else—a container, say—and is intended to then be composted after use, must have some mechanism for ensuring this happens. Clearly printed instructions, or an agreement with the receiving organization, or a method of collect-and-return are some possibilities.

To understand the circular economy with respect to any design process is to first understand the initial need of the product itself. A truly circular economy relies on not just a few adherents to the concept, but the whole global economy accepting and investing in it. Clearly this is unlikely to happen overnight, but all journeys must start somewhere, and the overbearing sense of urgency about climate change should be reason enough for everyone involved in any design project to place the need to make environmental concerns—environmental safety, if you like—at the very forefront of their thinking.

Environmental Impact—A Footnote

It is not the authors' intention to politicize environmental concerns: politics has no place in design. We must design the best product for the client within the remit of their statement of requirements and provide solutions that are right, proper, effective, and ethical every time. If in so doing, those designs are also elegant, creative, and sustainable then we have truly earned our spurs. But the world is driven by money and there is no single solution to the environmental disaster that awaits us if things do not change. But change will undoubtedly come.

It will come in the small things that many, many people do. Environmental concern is now much wider among the general public than even, say, five or ten years ago. Design started to consider environmental impact some time ago, and public attitudes to that impact has improved over that time, encouraging further designs to consider it. The Ford Motor Company, for example, launched the Sierra, the replacement to the much-loved Cortina, in 1982. It was the first mass-produced car to be specifically designed with aerodynamic efficiency in mind, rather than simply brand aesthetics. It was widely derided by both press and public, even being nicknamed the "jelly mould," and sales initially suffered, particularly in the UK. But aerodynamic efficiency is one way of reducing a car's fuel consumption and Ford's paradigm shift in car design was a bold and much-needed step forward. Today, we merely expect that a car has been designed as aerodynamically efficient as possible because we have become so used to the concept.

This is a further demonstration of the influence that design and use have over each other, as we discussed in the section on design influence: where design triggers a stimulus in the user by promoting a new approach that eventually becomes a natural component of their expectations, thus requiring it to be considered in all future designs of similar products (see Figure 3.5).

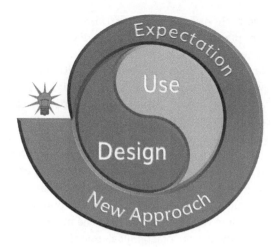

Figure 3.5 Influences.

Large organizations can often take bold steps when the rest of us are simply trying our best to do the job that we do. Very few of us will have the opportunity to change the world with a pen stroke. But if every design that we undertook made one small change to improve the sustainability of the end product we would all have had a hand in changing the world for the better.

Further Considerations

Provision of Materials and Manufacturing Techniques

After its discovery in 1825, aluminium was a rare and highly-valued metal. It is said that Napoleon made gifts of aluminium cutlery to his closest friends whereas his less-revered associates received cutlery made merely of gold. The invention of the first industrial production of aluminium in 1856, however, saw aluminium in the early 20th century become one of the most important construction materials, with production today exceeding all other non-ferrous metals combined. Endlessly recyclable, strong, lightweight, corrosion resistant—aluminium is invaluable in many industries such as aeronautics, construction, and transport. One important property of aluminium, however, is that its melting point is relatively very low compared to say, steel (660 °C compared to circa 1500 °C). Another, that in powdered form it is exothermic (meaning that fine particles created when machining aluminium can be explosible) can make it a difficult choice in certain circumstances.

Material engineering over the last few decades has become an extremely precise science. Metallurgy, and the creation of alloys for specific applications has long been understood, but even so there have been great strides in development in recent years. The development too of plastics since the discovery of celluloid in the mid-1800s has led to a world where, today, materials can be specified with very precise properties. Plastics technology has, however, seen a shift in popularity due to environmental concerns and there is now a move towards sustainable materials instead. Much progress is being made, for instance, with carbon-based materials such as carbon-fibre and graphene which, in particular, has many exciting potential applications.

In short, material selection in the design process is a critical decision, whether those materials are well-established or bespoke. And, as we have seen in the case of external influences, design and material choice can have complementary effects on one another. Returning to aluminium as an example, one key issue is that, unlike steel, the weld between two pieces of aluminium is weaker than the surrounding metal. This was one of the reasons that aluminium was not used more in the manufacture of motor car bodies until fairly recently. The design of chemical bonding agents, however, saw the increase of aluminium being used in the transport industry generally and in motor production in particular; from Audi's A8 model in 1995 (the first all-aluminium production car) to the new electric London taxi and van variant.

This example of advances in material use combined with novel manufacturing processes leads to the consideration of manufacturing itself. Again, technical advances over the years have led to such techniques as plasma cutting and subsequently, laser cutting, 3D printing, and 6-axis computer-controlled machining. The use of novel techniques and materials will invariably involve an increase in initial cost but these can often be offset against other considerations such as environmental impact, production time, or longevity. A good appreciation of the client's design intent will assist the designer in proposing such novel approaches, especially where there are few limits in terms of the ideas put forward for consideration. In proposing novel materials or techniques, it may be that the client sees an opportunity to resolve other elements of the design requirement, or even identify solutions to other issues that had not previously been subject to scrutiny.

We should remain wary of the effect of confirmation bias on the decision-making process as far as materials, equipment, and production processes are concerned. This tendency to digest information based on one's own preconceptions or beliefs can hamper the introduction of novel materials or processes that may, in many respects, be a better solution than those previously used. Confirmation bias can affect both the client and the designer equally, where both can be drawn too readily to solutions that fit with their respective predispositions, based on previously successful outcomes. This can often be to the detriment of the project, especially where novel or radical solutions could provide superior outcomes in terms of efficiency, safety, longevity, or reduced environmental impact.

Ergonomics and the Work Environment

We have discussed some of the legislation that applies to products in the workplace, and why designing with this legislation in mind promotes safety in general through the "safe to operate and operated safely" maxim.

There are, of course, many pieces of safety-related legislation that affect all aspects of how people interact with the products designed for their use. We can see, however, that all safety-related legislation is created to prevent harm to those interacting with these products and it is valuable, therefore, to consider some of the areas that, as designers, we should be paying attention to. It is also worthy to consider that in the profession of health and safety there is much current discussion about how there is too much emphasis on "safety" and not enough on the "health" aspect. The wellbeing of individuals and, increasingly, that which applies to mental health has gained much interest in recent years. Although perhaps predominately focused on the work environment and, moreover, the structures where this takes place, it should be considered when designing anything with which people interact, either at work, at home, or during recreation.

Space

We consider space not only by the space that the product occupies, but also the spatial environment in which it will exist. Designing a bespoke product to fit into a defined space is, of course, considerably more straightforward than designing a publicly-available product that could be used in a million homes across the globe. In the first instance, we can allow for operator and maintenance access as well as interaction with other products and the setting in general. With the latter, we must account for operator misuse as well as any number of environmental factors. Anyone who has knocked over a lampshade while playing a virtual reality game will understand.

Air Quality

Concerns for the quality of the air we breathe have been gathering pace for many years now. The smog that used to pervade across cities in the UK in the 1950s may be a thing of the past, eradicated by legislation brought about because of the disruption it caused, but today particulates in the air have been identified as the cause of many health disorders and are, similarly, being better monitored and controlled.

We should not only be considering the environment in which the product is to be used (products in dusty environments might require a level of ingress protection, or "IP" rating in accordance with BS EN 60529:1992 (BSI, 1992)) but also any emissions from the product itself. This may manifest itself as particulate emissions but also as electromagnetic radiation. This can ionize the air immediately around the product and can degrade air quality. Again, considering the potential operational environment of the product and the operatives' needs can help in formulating controls or countermeasures, such as fans, vents, exhaust ports, and so forth, as well as any relevant operational instructions.

For workspaces, there is increasing interest in air conditioning systems that have long been thought to be a factor in "sick building syndrome," where pollutants are carried around entire buildings causing ill health among the occupants. The pandemic of 2020 has also highlighted the effect that air conditioning can have on spreading viruses, and treatment systems, such as ultraviolet filtration, may now become more commonplace. Air quality is in fact a matter of legislation, with the Workplace (Health, Safety and Welfare) Regulations 1992 (HSE) requiring workspaces to be supplied with fresh air, as well as for systems for providing that air to be fitted with a warning device to let the operator know when it has ceased to function correctly.

Light—Quality, Quantity, Colour Temperature

Offices have come a long way from being lit by fluorescent tubes hiding behind dusty acrylic diffusers; their red-shift light having little effect on the computers of the day, with their black screens and green characters. Today's offices, factories, and warehouses can be floodlit with bright, diffuse light from LEDs with often selectable colour temperature settings. But control panel screens still often have quite narrow fields of view, which can cause the operator to miss vital information being displayed if they are not exactly face-on to the screen. Glare and reflections can cause similar issues and the wearing of spectacles with polarizing lenses can cause striping: the reason why airline pilots do not wear polarizing sunglasses.

But the level of light (lux) is not the only issue we must consider. The temperature (measured in Kelvin) can also affect operators trying to make quality judgments where the colour of something is critical. Normal daylight is roughly 5500/5600 K; lower temperatures are shifted red in the spectrum and higher temperatures are shifted blue. It seems that this knowledge continues to be ignored by fashion retailers as anyone who has bought an item of clothing in a shop, only to find it is a different colour when they unpack it at home, will know.

Stepped changes in levels of lighting can also yield poor operation. The lighting levels in a fire exit corridor, for example, need only be around 50 lux. A control panel half way down a dark corridor that has a bright white screen is unlikely to help the user to see properly as they turn away from the screen to resume their long, dark walk to the fire door. One solution, one that mobile telephone handsets use, is a simple lux sensor that automatically darkens the screen when the ambient light is low.

Green Spaces

Perhaps more for those designers in traditional construction projects is the consideration of green space. The calming effects of the colour green on humans has long been understood; it is the reason why hospitals, schools, and police stations often made extensive use of green paint on the walls. Today, there are such considerations as "living walls," which are essentially vertical plant pots, only on a much larger scale. Modern office spaces, particularly in large cities, often include green spaces in the rest and welfare areas in order to induce a natural, calming effect for the workers to enjoy.

The increasing awareness of mental wellbeing is leading to workplaces being created with more subtle and integrated natural tones than the odd pot plant dotted here and there. This can not only benefit the mindfulness of individuals but can also improve air quality too when designed to be integrated with air flow and conditioning systems.

Anthropometry

The study of human dimensions is possibly most elegantly demonstrated by Leonardo da Vinci's "Vitruvian Man" drawn around 1487, although it was not until the 20th century that it began to have industrial and commercial importance. Anthropometry is concerned with the *proportions* of the human body, and their correlation, rather than simple measurements, and we have today a large quantity of data due to many studies, including those conducted by NASA and the US military.

Anthropometric data are used to influence the design of such things as clothing, aircraft seats, and space suits. This information can be used to inform the size and shape of just about anything that a human holds, wears, or interacts with. It is closely associated with ergonomics in that, creating a workspace—an operator at a machine, for example—is concerned with both the range of movement the operator has to go through to operate the controls as well as the layout of those controls and their relative distances from each other and the operator. Aeroplane cockpits, although apparently complicated at first sight, are examples of how these sciences have been used to great effect.

Spatial Design

This fairly nascent discipline combines such specialisms as architecture, landscape design, and interior design to create a cohesion between buildings and those who occupy them. It is not to be confused with the "spatial coordination" stage of the RIBA Plan of Work (RIBA), which is concerned with the detailed information required to prepare a building for construction. Spatial design is another example of how once seemingly staid industries—in this case, construction—have realized the benefits of incorporating mindfulness into their core design requirements.

With traditional workspaces becoming ever-more challenged by such matters as homeworking and environmental concerns, as well as the possibly long-term effects of the global pandemic of 2020, it is likely that spatial design will become more important as the boundaries between work, home, and leisure spaces become more blurred.

Operating and Maintenance Procedures in Service

The methodology of operating and maintaining a product form as much a part of the design process as, say, the choice of materials. Additionally, *how* the product will be operated and maintained is as important as *who* will do the operating and maintaining. To demonstrate, let us consider our case studies and consider some of the "how" and "who" possibilities (see Table 3.4).

Table 3.4 Operation and Maintenance; Case Studies.

Example	How will it be operated and by whom?	How will it be maintained and by whom?
Nuclear power station	A power station will be operated to a rigorous and complex regime by highly-trained technicians. Every facet of the operation will be codified and strict hierarchies of control will be in place. There will always be present senior technicians to assume control in abnormal situations.	Maintenance procedures will follow equally rigorous systems and engineers will be very highly trained. Specific systems, such as permit to work, will be in place.
Office block	A wide range of operators from casual visitors to knowledgeable permanent staff (for example, fire marshals) are likely to frequent the building. Visitors may be unfamiliar with the layout and will rely on signage. Long-term users may enjoy discovering unauthorized routes or restricted access areas.	Building maintenance could be carried out by anyone from an in-house handyman to an ever-changing stream of contract engineers. The knowledge of control points, access areas, and building layout could therefore vary widely. Levels of skill in dealing with complex control systems could also be variable and instruction manuals may get lost, defaced, or ignored completely.
Warship	The operation of a warship is conducted by highly trained personnel with very defined areas of training and responsibility. There are also several layers of supervision and redundancy, or dual-role operatives. Operation of a warship needs to be considered against a potentially rapidly-changing dynamic environment.	Overall, maintenance on a warship is conducted to strict procedures by highly skilled engineers often when in dock. There may also be circumstances where running repairs and maintenance must be performed under possibly extreme conditions and using novel or imaginative solutions.
Home printer	The range of potential users varies between those with no practical knowledge of the product to those with advanced IT skills. The expectation of "plug and play" operability is high. The control interface should accommodate those with limitations in sensory perception and motor-function.	Products with no user serviceable parts should render access impractical. Access to consumables and possible paper jams should be straightforward and well signposted. There is the risk of people attempting unauthorized home maintenance, thus requiring consideration of protecting electrical components.
Motor car	Cars can be operated by users with a very wide range of driving skills, sensory perception, motor-function, and body size and shape. Added to this are the huge range of road types, weather conditions, altitudes, and cornering and braking forces that might apply. Additionally, cars can go through several iterations of ownership in their life as they become less valuable.	Maintainers of cars fall roughly into three categories: those trained by the manufacturer; those trained or skilled in maintaining any brand of car; and operator/maintainers with little or no formal training. Some safety-critical components (such as tyres and brakes) are generally the easiest to maintain or replace and non-manufacturer-specified components are widely available.

Whilst it may not be possible to design a product for every eventuality of operation, repair, or maintenance, it should be possible to *foresee* those circumstances that could impinge on its day-to-day use. This is not just the environment in which the product is used or maintained but also the manner in which it is used or maintained. As we can see, a product may have very defined modes of use with widely varying environmental factors (e.g., a car); or it may have a static presence with widely varying operators and even modes of operation (e.g., an office building). And, of course, every permutation in between.

The consideration of the foreseeable risks to the normal operation of a product is a key component of safe design; but, of course, not all risks can be avoided. Operation or instruction manuals are but one method of providing *control measures* to either risks we cannot avoid or operational limitations of the product itself. For example, the manufacturer of a motor car cannot prevent a future user of their vehicle fitting the wrong tyres, but the inclusion of the tyre specification in the owner's manual is at least an administrative control measure.

Training Provision

Both operational and maintenance modes of a product's life will, to some extent, be circumscribed by these operational limitations. As well as defining these limitations in operational or instruction manuals, it may also be necessary to provide training material too.

Training is a discipline as richly diverse and creative as perhaps design is and, indeed, the creation of a training programme or package is a design process in itself. It certainly centres around some of the same initial questions that design has to consider at the initial stages of the project: who is it for; how is it to be delivered and in what environment; what is it required to do and what are the foreseeable risks if it is not operated (or learnt!) properly? It should be clear from this that spending time assessing the training that would be necessary to carry out the proper operation of a product *at the early stages of the design* may well influence the design's maturity if it transpires that the level of training outweighs the product's usefulness.

Virtually every designed product will require the provision of some form of training in order for it to be correctly used and that level of training will, of course, depend greatly on the product itself. The level of training required to successfully operate a nuclear power plant will clearly be far higher than that for a motor car. Workers in a power station will undoubtedly undergo prolonged training programmes and operational manuals will most likely be long and complicated documents. Conversely, a car driver has received a standard level of training by taking a nationally approved driving test. The functionality of every car needs to be structured therefore within the parameters of that training standard: deviations are likely to cause errors in operation. The car manufacturer Saab, for example, made several models that required the driver to apply the handbrake before removing the ignition key, as the key barrel was fitted directly underneath the handbrake lever. This admirable safety feature unfortunately led to the key barrel often needing to be replaced in older cars as drivers had during the course of ownership wrenched the key out assuming there was an issue with it, rather than knowing the correct operation.

It may be that the creation of a training package or programme is, in itself, a design process and therefore the application of an effective design strategy is crucial. In this case, the client (or initiating need) for the design process is actually the product that requires a training package; but in all other respects the process for designing training is exactly the same as for the product itself. By conducting the effective management of designing a training package in this way, the various influences and risks can be assimilated and mitigated in the same way. Is there, for example, a risk that if people are not trained to the required level that some harm may occur? Or is there a potential, if the training is too involved or complex, that it may be ignored or adulterated? Is there, even, the possibility that suitably competent trainers will be available for the intended duration of the product's lifespan? Some of these factors may be answered by examining whether additional investment at the design stage of the product itself might mitigate the need for complex or overburdensome training at all: a question of capital expenditure (CAPEX) being balanced against operational expenditure (OPEX)—a question surely that invites comment from the financial department.

Obsolescence

Obsolescence might normally be thought of as when a product is impacted when one or more components are no longer available in the supply chain. In fact, if we consider obsolescence as the inability of the product to continue to function in line with its original intent then we find there are several causes for it. All of these causes require suitable management whether during the design process or the life cycle of the product. Some causes for obsolescence may be:

- a change of components;
- the sourcing of raw materials;
- the supply of spares or consumables;
- the availability of test equipment;
- levels or standards of training;
- changes in legislation.

The components required in any product design will themselves not only be subject to their own design process but also to many of the external factors we have discussed, including that of obsolescence. Good supply chain management is an obvious way to control this issue and certainly proprietary components may well be able to be sourced from different manufacturers. Bespoke, novel, or unorthodox components, however, may present more of a challenge in securing supplies.

The supply of raw materials should be thought through at the design stage as an external influencing factor. The demand for battery-powered electrification of many products today is having a huge impact on the global supply of some exotic materials such as lithium, cobalt, and manganese, for example. This would certainly warrant being recorded as a risk in the initial stages of the design process.

Spares and consumables are, essentially, components forming part of the original design and, therefore, can be subject to the same forces of obsolescence. Similarly, of course, there may be after-market suppliers springing-up to fill any voids in the supply chain, but this cannot be relied upon at the design stage. As with components, the supply of spares and consumables can be managed through the examination of the capability and capacity of the supplier(s); what we might consider to be part of their overall *competency*.

In consideration of these first three factors, it should be noted that the ideal of the circular economy may present additional threats *and* opportunities in the supply chain. The notion of a product designed to be refurbished during its life and then re-sold will most likely suffer from an initial shortage of spares but will, once enough products are available in the market, hopefully reduce the reliance on the supply of raw materials. Additionally in this case, a robust "return-to-base" logistics set up will be required if the product is to be successful in its ambitions.

Available test equipment might become an issue in connection with a particularly complex or novel product that requires a specific type of testing. The testing equipment may suffer from obsolescence of components or spares in exactly the same way as the product, causing issues for its maintenance or repair.

Levels and standards of training may not appear to be a particularly obvious cause of obsolescence but even well-established training methodologies can be subject to change over time. Consider the changes to the driving test in the UK made in 2017, or the successive iterations to the Institute of Engineering and Technology (IET) Wiring Regulations, which create the need for re-training in electrical works. Although it may not be possible to foresee changes to training, clearly the more bespoke any training is the more likely the risk that it may change over time in line with technology and progress.

Legislative changes can introduce obsolescence in any of the preceding causes we have discussed but usually happen over protracted periods of time. This would normally be foreseeable and would be recorded as a risk at the initial stages of the design process. Changes in overseas legislation and changes due to emerging events, however, can cause issues in the supply chain with very little warning. The Firearms (Amendment) (No.2) Act 1997 is an example (admittedly an extreme one) where an entire industry and supply chain were essentially made illegal almost overnight through the banning of owning and supplying an array of hitherto legal weapons. Although the legislation was rightly in response to the appalling tragedy at Dunblane in 1996, it is an example of how sometimes legislators can react relatively quickly to dramatic events that subsequently has wide-ranging consequences.

Keeping track of these myriad causes of obsolescence may require anything from simply managing the risk register to employing an obsolescence specialist. For products that require a supply of components or materials, this can often be managed through the supply chain, and may even form part of the contractual obligations put in place by the procurement department during the project. Hence the reason it can be beneficial to engage with the procurement team at the early stages of the project in order that they establish for themselves the impact of failures within the supply chain to the end product. This is particularly the case where the product relies on bespoke, unorthodox, or cutting-edge materials, components, or technology. An example can be found in the case of the Tesla electric car company, which built its own vast battery factory in order to maintain a suitable supply of lithium batteries, as well as to conduct research and development on new battery technologies that could then be integrated effortlessly into the design of new vehicles.

Influences Surrounding the Product Life Cycle

We have discussed in this chapter some of the factors which, during the use of a product, may have an influence on the design process and the reasons for this influence. They are as follows:

- Maintenance: service records can highlight issues with reliability or access, as well as the planned timing of service intervals in comparison with actual in-the-field outcomes.
- Usage data: records of usage, in terms of runtime, function, and environment, can provide valuable information about how the product was actually used compared to its design intent.
- Health and safety risk assessments: how the use of the product, evaluated by occupational safety practitioners, can highlight potential safety concerns with its operability.
- Regulatory: changes in regulations affecting the product may well affect future products of the same or similar design.
- Obsolescence: this may come about either due to materials or components becoming obsolete or because of the product's obsolescence due to societal impacts, or simply changes in the client's requirements.

We also discussed how warranty issues with previous products could be drawn on to propose alterations in future designs of similar products. Such issues may not be simply from general product recalls, or customers returning faulty goods. Warranty issues may also include bespoke products that have failed, sometimes catastrophically. Where a bespoke product, irrespective of size or complexity, is designed and produced to a specification, it is *warrantied* to operate to that specification by the designer, whether explicitly or implicitly. Failure of the product in this way is a result of one or more of the following:

- The business case not fully appreciating the necessary implications and considerations of the project.
- The statement of requirements not being adequately framed around the actual need of the client.
- Consideration of external factors—use, maintenance, regulatory, environmental, and so forth—not being properly undertaken and accounted for.
- The designer not adhering to the statement of requirements.
- Inherent risks in the project not being properly identified, understood, or controlled.
- The producer not following the final design correctly.
- Changes to the specification of the design, materials, or components without proper consideration of the implications to the overall design intent and specification.
- Operational and maintenance procedures not being adhered to correctly or not being adequately documented.

It may be seen from this list that there are many reasons that a product can fail during its lifetime of use and that, equally, all of these events might be easily prevented through effectively managing the project from the very outset of the design process. An effective design strategy does not begin with the designer lifting pen to paper; nor does it end when the producer assumes responsibility for producing the designed output. Rather, the design process must encompass all factors and influences throughout the product's expected lifetime in order to adequately identify and make provision for all the salient factors that may impinge upon the product, its use, maintenance, and disposal. It is only by implementing this effective strategy that the safety of the product, the project—and, most importantly, all those who interact with the product—can be assured as far as reasonably practicable.

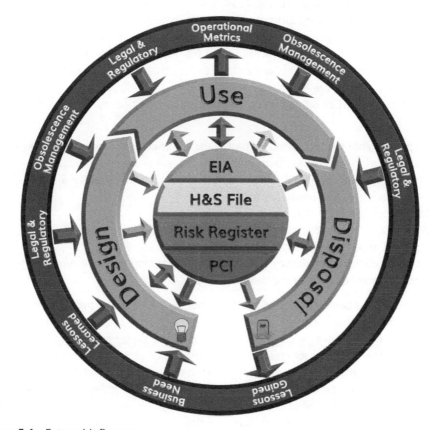

Figure 3.6 External Influences.

The external influences, information, and safety factors that affect a product's entire life cycle share an interconnectedness with the design process (see Figure 3.6). This is because of the varied governing elements that must be considered and managed during that process: regulatory requirements such as the health and safety file; mandatory requirements such as BIM; and the requirements of relevant standards. There are also elements that are influenced by the process, such as operational data or metrics, and any lessons learned. All of these influences must be considered, from the "lightbulb" moment of recognition that a product is required by the client, to the point where the product is finally disposed of.

It should be noted that this illustration is for just one iteration of the product's life cycle, meaning that where the product is varied, extended, or repurposed then the cycle begins again. Any of these events is a change to the original design intent and must therefore be effectively managed and documented in order to ensure that the variation, extension, or repurposing is safe within the parameters of the new design intent.

Managing/Maintaining the Design Objective

As we have discussed, the design of any product must be constantly measured against the client's needs as captured in the statement of requirements. This may be refined and challenged as the design matures, particularly as emerging risks are identified. The need to continually measure the design is assisted greatly by conducting reviews at appropriate points in the design process and, of course, with the appropriate stakeholders present at those reviews. This will help to ensure that the design remains on track and that errors later on are minimized.

Some of the reasons or influences on why the design solution may vary as it matures might be:

- the imposition of new regulations, either in the originating or destination market;
- regulations being updated, perhaps due to an incident or socio-political event;
- advances in technology affecting materials, techniques, or market demands;
- supply chain issues and obsolescence;
- socio-economic impacts that affect market demands;
- changes to operational requirements for the product or its functionality;
- geopolitical implications, such as border disputes or trade wars;
- client prevarication or hesitancy, which prevents clear, timely decision-making;
- poor overall management of the design process.

Whilst the design is being developed and under the control of the designer, there may be some tolerable variations that will allow it to continue to meet with the client's expectations. Tolerance to the design can be integrated into the statement of requirements to allow the designer the room for creativity during the design process. Again, understanding the initial need for the product will greatly assist in deciding how much tolerance can be factored in, if at all. In the case of our home printer, for example, must it have, say, twenty pages per minute printing speed in order to be a market leader or can it run a little slower in order to make use of a readily-available gearing mechanism that will reduce costs? Managing the design intent is about navigating the design through the many influencing factors, within the tolerances of the statement of requirements and all the while meeting — or exceeding — the client's expectations (see Figure 3.7). Designing a home printer with an in-built coffee machine may be a market first, but is unlikely to be what the client envisaged.

This may not be as flippant as it sounds when one considers the efforts of a chemical researcher working at the American conglomerate 3M who was trying to develop ever stronger and better adhesives. What he ended up discovering was "microspheres," which were sticky but did not bond permanently — considerably outside of the client's original intention. That adhesive, however, went on to be used on Post-It Notes™, which proved to be a global success for the company.

Once the design is produced and validated and the product enters operational use (in service), the designer may still retain an influence, to a greater or lesser extent, over the

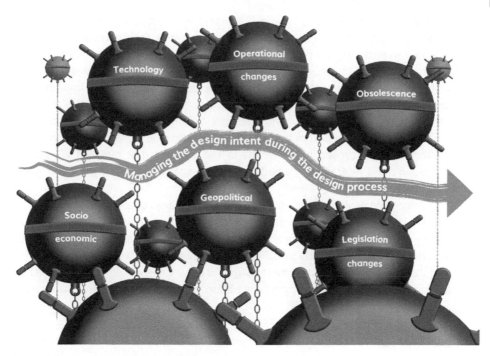

Figure 3.7 Managing the Design Objectives.

maintenance of the design objective. This could be due to: being established as the design authority; having enduring responsibility from a contractual or legal obligation (such as CE marking); an emergent requirement for a design change; or through upgrades to the design during the product's in-service use.

Being established as the design authority for a product during its serviceable life is usually reserved for those projects involving large, complex, or very high-value products. It is where the designer retains an interest in the product's service life due to its importance and may be because the designer or design team is also the client—as is sometimes the case with a power company building a power station—or they may be embedded contractually with the client—as in the case of a warship where a defence contractor builds it on behalf of a nation's navy.

Maintaining the design objective due to a legal or contractual obligation may arise if the product fails to deliver on its functionality through some failure or error in the design. This is why it is so important to test any design against as many risks as possible. A home printer or, perhaps, a washing machine that does not meet its projected lifespan, or fails catastrophically during it, is an example of where redesign is the responsibility of the original designer even if it proves to be an issue with the specification of components. In these cases, it may be readily seen that the designer can protect themselves by ensuring all the decisions, and the communications surrounding them, are recorded appropriately. It is also an example of how use can affect design and vice versa and also why it is important to understand the operational environment of the product as well as issuing robust instructional material for when it enters service.

Sometimes, products go wrong due to some design error. Although the majority of errors are caused by such avoidable matters as cost-cutting, not identifying salient risks and failing to appreciate all of the influencing factors, we must accept that occasionally mistakes are made. It is in these situations that the designer clearly owes a professional debt of honour in rectifying the problem, again within the original design objectives. An example of a design that suffered from a possibly unforeseen design error is that of 20 Fenchurch Street, London, also known as the "Walkie Talkie." The 38-storey building was designed to curve outwards as it rose resulting in a slight parabolic effect which, when combined with the fully-glazed exterior, became a concave mirror on the southern façade reflecting the sun for part of the day. Consequently there were reports of vehicles and buildings being scorched in the street to the south as the building effectively focused the sun's rays. Interestingly, a building in Las Vegas designed by the same architect suffered from a similar problem.

Often in compound designs, for example, those where many components are designed to form a whole, there is a requirement to upgrade certain elements of the design due to any of the reasons listed above or simply through some aesthetic desire. The motor manufacturing industry frequently modifies and upgrades its models through their lifetimes and these alterations must all be managed to ensure the whole design retains its original objectives in terms of aerodynamics, performance, driver experience, and so on.

Design Management—Summary

We have seen that the design process should not be viewed within the restricted environs of solely the technical response to the intended design. The many considerations and external influences that must be brought into the mind of the designer are accompanied by further considerations of the life cycle of the product. Ensuring the product meets the requirements of the client, as well as the expectations not only of the regulatory environment but also those of the intended market are allied to the critical importance of the global environment in which we all have a part to play. All of these influences carry both threats and opportunities that should be evaluated, recorded, and reviewed because it is the effective understanding of risk that shapes the client's true requirements.

It is not the intention of this book to suggest that all designers should acquaint themselves with the entire projected life cycle of all their designed output, but in considering the output's environment and operational use, which we have already discussed as being necessary considerations, we have begun to open up that life cycle to scrutiny. "Management" is the process of controlling things or people, and the management of a project is exactly the same as the management of a design process. It begins with planning and continues through controlling its progress. It is said that failing to plan is planning to fail and this is never truer than in design management.

Identifying the needs of the client and transcribing these into a firm statement of requirements is crucial to providing a clear framework for the designer to create within. Registering the many external factors that can influence a design will add colour to the designer's palette, allowing them to create appropriate solutions within the frames of reference within which the product will function. And understanding the requirements of the people who will interact with the product, and their environment, will give the product form and texture. Greater planning at the beginning of any endeavour will almost always yield dividends later on in terms of cost benefits, time savings, and error reductions.

4

The Management of Risk

The Importance of Managing Risk

Undoubtedly, the most important part of any design process is the proper management of risk throughout. This may at first appear an unusual statement, but the paramount reason for the design process is to provide an output that satisfies the client's requirement. That requirement, howsoever it should be ultimately formed, *must* be safe to use. It must be safe to use because safety legislation is at the very heart of virtually all legislation, not just in the United Kingdom but across the European Union and much of the rest of the world. By demanding safe products to use, and safe places to work and live in, our society has placed its trust and the responsibility of safe design firmly in the hands of designers.

We have, therefore, a legal duty to ensure that the products we design are safe to use, work with, and operate; a legal duty that is then transposed onto the client in the use, operation, and maintenance of the product. But we also have a moral and ethical responsibility too. Ethical because we owe a duty of care to our fellow humans to ensure and preserve their safety; moral because it is what society has ended up demanding after many years of developing its affinity for safety in all areas. And central to this tenet is the appropriate management of risk.

As we have seen, design forms just one part of a product's entire expected life and that both the design and the use of a product are interconnected so that each affects the other. The design of something will affect the way in which it is operated safely, and the use of something may affect the way future products, or additions to them, are designed.

During the design stage, most of the risks will tend to be foreseeable, in that there is a certain amount of supposition required to identify the risk. During service or use, there will be operational experience and data collected that will enable better understanding of risk. The risks during this latter stage will tend, therefore, to be more observable.

As an example, during the 1960s space programme undertaken by the National Aeronautical Space Administration (NASA) in the USA, one of the many considerations of risk that were *foreseeable* was that of the astronauts falling ill with food poisoning. NASA decided it needed a method of ensuring that any food prepared for the astronauts

An Effective Strategy for Safe Design in Engineering and Construction, First Edition.
David England & Dr Andy Painting.
© 2022 John Wiley & Sons Ltd. Published 2022 by John Wiley & Sons Ltd.

during their spaceflight was prepared under the most stringent of conditions. This need led to a separate design project (the client in this case being the space programme itself) to develop a system of ensuring the reliable quality of prepared food; the output (product) of which was the Hazard And Critical Control Point process. This process, commonly referred to as HACCP (pronounced "hassup"), is widely used today in commercial food factories, warehouses, and restaurants. During the process, risks to the quality of the food produced become *observable*: for example, chilled food being stored above 5 °C or defrosted food being re-frozen, and so forth.

Risk is measured by quantifying the levels of severity and likelihood and the control measures needed to be put in place should be commensurate with these. Clearly, during operational use, the knowledge of risks with a product should benefit from the experience and data we have acknowledged previously. This should allow control measures to be more accurately relatable to the risk involved. During the design stage—where risks tend to be more foreseeable, and therefore requiring a certain amount of estimation—control measures may be less accurately targeted, or more expensive to implement, or less successful in their outcome. Or, possibly, all three. This is why the designer will be greatly assisted by having as clear an understanding as possible of the product's intended use, environment of use, and the statement of requirements, as well as an understanding of the client's *initiating need* for the product.

To demonstrate this, let us consider one of our case studies, the office block. We are to build a new office block in a location where the access route has power transmission lines strung above it. This will have been identified in the pre-construction information and is something that we note in our design risk assessment. The client has requested that the design of the building incorporates as many prefabricated elements as possible to speed up the process of construction but the presence of the transmission lines may affect the size of the elements that we are able to bring to the site. It may also hinder the size of the crane that we are able to bring on site, or the ability of the crane to deploy correctly or reach all of the site effectively. It can be seen that the client's requirements, the size and design of the structural elements, the overhead power lines, and the type of crane are all inextricably linked: the adjustment of one is likely to affect the others.

Risk Management Process

The management of risk is its effective identification, assessment, and mitigation, and begins with establishing the context of the risk: financial, reputational, safety, and so forth. Assessing risk relies on deciding who or what is at risk, evaluating or prioritizing the mitigation or control measures available and then implementing them. Following this there should be a mechanism for monitoring and reviewing those measures to ensure that the risk remains controlled within the context established previously. With a financial risk, this might be to reduce the level of exposure to financial loss to a value set by the organization. With health and safety risks this would be to a level as low as reasonably practicable as defined in health and safety legislation. The British Standard on Risk Management, BS ISO 31000 (Risk management. Guidelines, 2018) portrays this process with the important addition of a communication and consultation subprocess throughout (see Figure 4.1).

Figure 4.1 General Risk Management Process.

This subprocess allows for all relevant stakeholders—senior management, depart-mental management, operatives, maintainers, and so forth—to be engaged in all stages of the management of risk. This can be especially useful in identifying appropriate con-trol measures and in ensuring that they are properly implemented. In health and safety legislation, this consultative approach is in fact a stipulation of the Management of Health and Safety at Work Regulations 1999.

The process of risk management is straightforward but as with many professional disciplines the results can depend entirely on the quality of the investigation and the translation of the results. Let us examine this process in more detail to understand the objectives of each step and the supporting processes behind them (see Table 4.1).

Table 4.1 Risk Management Objectives.

Stage	Notes	Supporting processes
Identify	The context of the environment from which the risk stems must be known	This comes from professional knowledge of the sector or industry in which the design will form an output.
	The objectives of the environment from which the risk stems must be understood	This can be gained from appreciating the client's and other relevant stakeholders' requirements.
	The risk appetite of the organization should be understood	This can be determined from the client's management structure, quality policy, document control, and safety culture as well as their corporate strategy.

(Continued)

Table 4.1 (Continued)

Stage	Notes	Supporting processes
Assess	Identify the risks	Complete a thorough design risk assessment by looking at the strategic, financial, operational, and maintenance risks to the project as well as any potential emerging risks that may affect it in the future.
	Deciding who or what is at risk	The design risk assessment should examine how the identified risks will affect the client as well as those connected with operating, maintaining, and disposing of the product. Also examine risks to the product itself such as failure, obsolescence, fire, and environmental impact.
Prioritize	Evaluating the type and level of risk	Establish if risks pose a threat or an opportunity. Consider this in conjunction with the client's risk appetite. Rate the risks in accordance with the potential severity of impact posed by each one. Align this rating with the client's risk appetite, remembering that low probability risks can sometimes have the greatest severity of impact.
Mitigate	Apply the general principles of prevention	Eliminate risks at source: designing-out risk is the simplest solution to promoting safety throughout the product's life.
		Reduce remaining risks: using design engineering or by replacing the hazardous with non- or less hazardous.
		Control residual risks: by providing protection systems as well as operational information, instruction, or training.
		Controls necessary for the safe use of the product can be delivered in instruction manuals, health and safety files, residual risk registers.
Review	Recording and reviewing the assessment	Use a formal design risk assessment to record the findings. Align this with the project's risk register, or the client's own risk register if they are to accept responsibility for the product on completion. Update the register(s) regularly, especially where emerging risks have been identified or where risks arise during the design, production, or in-service operational stages.

The Risk Register

The risk register is where all identified risks to any project or undertaking can be grouped and recorded centrally in order that they can be discussed and reviewed by relevant stakeholders. Risks should be grouped according to their particular category of influence: for example, financial; operational; administrative; supply chain; and so forth. This way, each group of risks can be reviewed separately from the main register by a relevant group of stakeholders. The procurement department might, for example, take the "supply chain" section away to discuss with their suppliers any salient risks that they may be able to resolve.

A risk register is similar in construct to a risk assessment but differs in one important criterion: a risk assessment is generally designed to identify risks and their relevant control measures for a particular activity—a workplace function, for example. A risk register is a much broader list of risks that can affect a project, a design process, or even an entire organization. A risk register can also deal with more ethereal risks than a straightforward assessment and this can mean that for some risks there are no identifiable control measures currently able to be put in place. For example, a change in regulations in an overseas market may cause an organization a great deal of work to adapt to, but without knowing any details of the change, it would be impractical to institute a precise strategy to deal with it. This risk remains, however, and should still be recorded in the register.

A register can also be used to prioritize risks to ensure that those that create the greatest or most imminent threat can be handled accordingly. This may include assigning a "champion" to take responsibility for implementing any necessary control measures, allowing identification of leading decision makers in a project's communication management plan. Project registers would ordinarily be closed at the end of the project, perhaps after any "lessons learned" review in order to capture any risks that could have been dealt with more appropriately or in a more timely way. Registers used by organizations invariably remain "live" indefinitely to allow them to be constantly reviewed and updated to include emerging risks, improved control measures, or revised champion's details.

Careful and dispassionate analysis of the risks to the design risks as well as to the project is vital in ensuring that observable, foreseeable, and emerging risks are identified and suitably planned for.

Influences on Risk Management

The British Standards Institute, in the standard BS ISO 31000:2009, state that:

> "*Organisations of all types and sizes face internal and external factors and influences that make it uncertain whether and when they will achieve their objectives. The effect this uncertainty has on an organisation's objectives is 'risk'.*"

In other words, risk sits at the intersection between one's objectives and the uncertainty of achieving them. Risk can therefore manifest itself as both an opportunity or a threat, and an organization's (or an individual's!) willingness or otherwise to engage with it is their *risk appetite*. We have seen the importance of understanding the client's intention for the design, as well as the micro-intentions of the stakeholders engaged with it, and this understanding helps to elucidate our perception of the appetite for risk that the client's organization, as a whole, has for the project overall.

In the identification of risk, we should be aware of the effect of both confirmation bias and normalcy bias. These are common behavioural traits in many of us and manifest themselves in much of our daily lives. Confirmation bias is the predisposition of a person to interpret information in a manner that reinforces their own preconceptions or beliefs. An architect who may have studied the brutalist school of design might tend towards proposing this to any client offering them a commission. As only those clients who like the proposed design will actually engage the architect (those who do not enjoy the brutalist look will clearly not), then the architect might well feel justified in continuing to propose such designs as all their clients evidently seem to like them.

In risk management, this bias can manifest itself in only identifying risks with which one is familiar, or in failing to accept plausible risks altogether.

Normalcy bias is a form of behaviour that can prove to have perilous consequences. It is a person's refusal to plan for, or react to, a threat of which they have no previous experience. It is a form of denial that the threat, and its impending potentially disastrous outcome, has a relevance to the here and now. In risk management terms this may manifest itself in a person's refusal to accept severe but infrequent risks as requiring any form of mitigation. The owner of a hotel may not have experienced a fire before and may therefore be reticent to the designer's proposal of an expensive water fogging system, which would not only contain a fire, preventing structural loss, but would also minimize water damage to the surrounding area.

The biases of the organization identifying and managing risk, or the individual doing so on the organization's behalf, form, when combined with the culture and corporate behaviour of that organization, its risk appetite. That is to say, the amount of risk the organization is prepared to accept as tolerable. Organizations that are risk averse will tend to shy away from even moderate risks whereas those with large risk appetites will engage readily with it, perhaps even encouraging it. One of the issues that led to the global financial crash of 2008 was that banks, who by tradition had been fairly risk averse, had engaged in activities that, although leading to high financial gains, also carried substantial risks of failure. Although the acquisitive part of the culture of certain banks was to relish the potential profit, it transpired that the culture was perhaps not mature enough to deal properly with the risk management aspect.

Aside from the behavioural traits and culture that help to make up an organization's risk appetite, there are several other factors that influence how the management of risk can be conducted (see Figure 4.2). The adequacy of the risk assessment process is fundamental to this in that, risks to the organization, project, or venture must be properly and thoroughly identified. If they are not, then clearly mitigation or control measures may not be put into place, or they may be inadequate, inappropriate, or ill-timed.

It is also important to understand the environmental context of both the organization and the risks, with regard perhaps to the industry sector or regulatory environment in which the organization operates. When the UK began the process of leaving the European Union in 2016, for example, there was very little direct impact on business; the agricultural sector suffered a loss of adequate workers, however, due to the uncertainty among migrant workers from Europe who made up the bulk of that sector's workforce.

Figure 4.2 Influences on Risk Management.

Environmental constraints will impact on risk through both those of a local and global nature. Local constraints might be those concerned with the immediacy of the organization's location: a venture to greatly expand a factory's output might be hampered if it sits within a crowded city centre for instance. On a grander scale, the concern for the planetary environment gathers pace all the time, and rightly so. Risks must be managed within the constraints that industries, governments, and societies are increasingly demanding.

Ethical constraints are akin to those of the organization's culture and appetite. There are certain expectations in some industries and professions, created either by the industry itself or by societal expectations due to perhaps historical or cultural reasons. We might return to the global financial crash of 2008 and ask if the propagation of high-risk financial instruments could be deemed ethical on the part of the banks considering the large amount of trust placed in them.

Risk Appetite

The Institute of Risk Management defines risk appetite as "the amount and type of risk that an organization is willing to take in order to meet their strategic objectives" (Institute of Risk Management, NK). In other words, it is the level of risk an organization is prepared to pursue. The institute suggests risk tolerance, however, is what an organization is prepared to deal with. In the profession of risk management these two terms have clear, if very closely-related connotations then; but for the purposes of this book, we shall treat them as meaning the same thing. That is to say, what levels of risk the client organization is prepared to accept as *tolerable* within its normal operational parameters.

We all have our own innate sense of what is, and what is not, an acceptable level of risk and this can be influenced by things such as age, upbringing, previous experience of threats and opportunities, and so on. Similarly, the risk appetite for an organization is not a single, fixed concept and may be quite a complex calculation for a risk manager to determine. However, the overriding *sense* of an organization's appetite for risk may be gauged through simple conversation, perhaps with reference to the type and extent of conceptual solutions that it finds acceptable in response to an initial idea. In this regard, it is similar to gauging an organization's safety culture through observation, questioning, and inference.

Linked to risk appetite are the different ways, or strategies, of dealing with risks. Understanding these will help to inform the risk register, where it can be decided how each risk (as either a threat or opportunity) is to be dealt with according to the level of risk appetite. There are four risk strategies that can be deployed (see Figure 4.3).

Figure 4.3 Risk Strategies.

Avoidance. By avoiding risk, we eliminate any potential harmful (or positive) outcome. This is the first iteration of the general principles of prevention and may often require an alternative project or design strategy, which may obviously have ramifications for the core constraints of time, cost, and quality.

Avoidance provides the greatest benefits for potentially the highest cost impact to the project.

Transfer. This is another way for the risk owner to avoid risk, but in this case the risk is transferred to another party. A simple example of this is in the purchasing of insurance for a project. Here, the risk—financial in this case—is not removed but its ownership is exchanged for a premium. Another example might be an engineering company that occasionally machines exotic metals: they may choose to have an external company perform any required machining to specification instead of performing it in-house.

Transfer offers high levels of risk avoidance at a relative cost to the project.

Mitigation. This is the route that is most usually taken in any form of design. The iden-tified risks will have suitable control measures put in place, again following the general principles of prevention, in order to reduce either the likelihood or severity, or both, of the threat being realized. Control measures may be anything from straightforward engi-neering controls to information, instruction, and training regimes.

Mitigation can provide good levels of risk control at a relative cost to the project.

Acceptance. If the risk has a low likelihood but high severity, or if the risk owner feels that existing systems in place confer a good level of risk control already, then they may choose to do nothing and accept the level of risk. In health and safety terms, acceptance of risk is not a relevant risk strategy but, in project terms, the client may well accept a particular threat without further intervention. A client may request that the design of a dock does not include provision for changes in sea levels over a 50-year projection, for example.

Acceptance offers no risk control and is usually without cost to the project.

External Influencing Factors

We discussed how design does not operate within its own isolated sphere of function. We looked at the external factors that can affect design and, conversely, how they can be affected by a design output. A product is clearly designed to be used and that use can be affected by perhaps some group or body to whom the users belong. That group or body will no doubt be in some way representative of a form of industry who, in turn, may have their own ideas and influences brought to bear on the design/use process. Outside of the industry, of course, is society in general with all the relevant concerns and influ-ences of its own. The design of a novel product solution can, for example, become so widely accepted through its use that, eventually, through acceptance by industry repre-sentatives and bodies, it becomes a societal expectation, thus affecting how all similar products are designed in the future (see Figure 4.4).

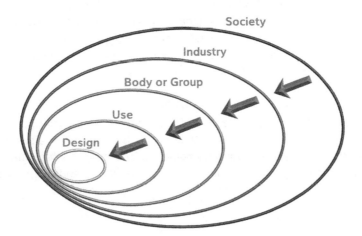

Figure 4.4 External Influences on Design.

In the preliminaries to beginning a design project, the requirements and the operational environment will have been established. These should then be examined in the risk register for potential risk outcomes. These outcomes may be determined as either threats or opportunities, which may be influenced by any of the wider contexts in which the project sits. Let us look at an example that relates to how the design and subsequent use of a building could affect each other with both positive and negative influences (see Figure 4.5).

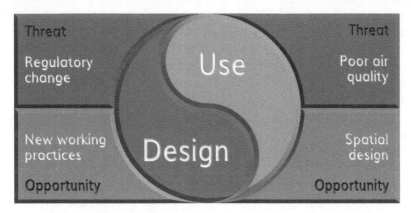

Figure 4.5 Threats and Opportunities.

Imagine that a negative influence has been identified during the design stage due to an upcoming regulatory change that will require far more robust fire detection and prevention measures to be included in this type of building. A positive influence has also been identified in that the client's organization wishes to implement new working practices that will mean that fewer work stations will be required, perhaps because more people are working from home. Both of these influences on the design will affect the way the building is used in the future.

Imagine now that during the use of similar buildings, a negative influence has been identified in that poor airflow, both in terms of the quality and quantity of air, has caused the buildings' occupants to suffer from a general malaise and low-level sickness. Equally, a positive influence has been identified in that the redesign of office workspaces, which included better spatial design and aesthetic quality, has improved the productivity and mental wellbeing of the staff. Both of these influences during use can affect the design of similar buildings in the future.

The situations in this example can only be identified by a designer who, firstly, has rigorously analysed the client's statement of requirements and, secondly, has the relevant skills, knowledge, and experience in their respective field. It may also benefit from the engagement of all the relevant stakeholders in the project, which might include operatives, maintenance personnel, human resources managers, and safety practitioners.

It should not be thought incumbent on the designer to interview all of these individuals or groups during the preliminary stages of the project, but it should be borne in mind by the designer that the views of all these stakeholders (for they will certainly become stakeholders in the finished product) should be reflected somewhere in the design.

There are many external influences that may affect any design project. The extent to which each, or indeed any, of them will have an effect, and to what degree, is entirely dependent on the peculiarities of the project itself. In assessing any risk, it can sometimes be challenging to bring certain items to the attention of all the relevant stakeholders, especially with those risks that may be perceived by some as having a very tentative effect on the project. The discipline of risk management, however, should not treat any risk with disdain; it merely applies levels of likelihood and severity to each risk in order that appropriate measures can be taken as required. By doing so, the management of risk can be coordinated to implement these measures in an objective manner: that is to say, that immediate risks, or those risks that pose the greatest perceived threat, are dealt with more adroitly than those perceived as having a less severe impact or likelihood of occurrence.

In essence, no risk is considered "off-limits" in risk management but some clearly require a higher priority of attention than others. Additionally, there are risks whose control measures can be implemented quickly due to their relative ease of introduction, whereas other risks may require large-scale investment or changes to processes. An example of this is the Windscale nuclear plant in Cumbria, UK, where nuclear waste from the Calder Hall power station was stored from the 1950s onwards. It was always known that something would have to be done about the waste produced by the power station but at that time the technology did not exist. It wasn't until the 1980s that the Thermal Oxide Reprocessing Plant was able to recycle nuclear waste, not only from the UK but also from all around the world. This was due to huge investment and advances in technology and materials handling. An example of the function of the time−cost−quality balance: in order to provide the *quality* of nuclear waste reprocessing necessary, a large amount of *time* and *money* were required.

It cannot be advocated that any designer should shoulder the burden of examining every risk that may or may not affect any given project. Nor should every design require adapting to every potential mitigation measure. But one should be aware of the potentiality of risk, and take corrective measures as necessary. This may simply be a case of enquiring if any particular risk has been considered by the client, even if the response is less than gracious. Planning early on for given eventualities, even those that may seem remote, will always save wasted effort and cost in the future. The resultant socio-economic fallout of the global coronavirus pandemic in 2020 meant that designers began to consider the layout of future workspaces so that social distancing measures could be more readily implemented in case of any further outbreaks. A pandemic would have featured high on a risk manager's register, as it surely was more a matter of "when, not if" it would happen. The treatment of workspace design though had rarely considered this "wildcard" of a risk, and so office spaces had to be quickly reorganized to accommodate new, distanced methods of working. This lack of consideration for an extreme and yet seemingly rarefied event is a trait of normalcy bias. Unfortunately, it is often the case that rare events can have the greatest impact: what we shall discover are termed "black swan" events.

At this current time in history, governments around the world are well aware of the threat of climate change and yet there remain contemporary designs that do not appear to adequately consider this. Changes in tidal height, for example, were often considered over a period of 100 years for infrastructure designs near to the sea. It is now not uncommon for this consideration to be 50 years or even less. Environmental factors are but one of several categories of risk for which any design may have to account for (see Figure 4.6). Each of the categories listed here is accompanied by two examples of how

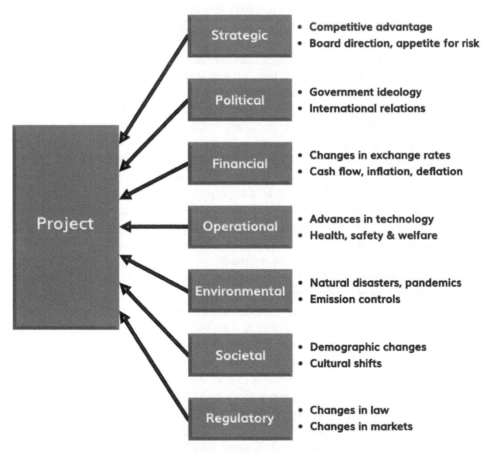

Figure 4.6 Example Influencing Factors.

that category of risk could be realized. The client, of course, may take ownership of some or many of these risks, depending on their risk appetite, and therefore the designer may be exonerated from considering them in the design process. This should be documented accordingly in the design risk assessment to demonstrate the verisimilitude of the decision and the authority that lay behind it. This is where a design risk assessment or risk register is important in not only being created but suitably updated, in addition to properly minuted meetings and the management of project documentation.

The global pandemic of the coronavirus SARS-CoV-2, which resulted in many hundreds of thousands of deaths worldwide and caused widespread financial downturn in some national economies was not, as may be imagined, an emerging risk. The imminence of a global pandemic had been understood for many years but regrettably had been all but ignored by many governments and global agencies. The financial fallout of a pandemic is, however, an emerging risk, as the world struggles to reinvest, and financial practices that have been long understood and practised become potentially passé in a new world order.

Clearly, not all designs will be subject to emerging risk; but it can be surprising how many may be influenced to some degree. As part of the initial recording of risk at the beginning of any project, the possibility of exposure to emerging risk should be identified and documented. Allowing the design to "flex" in order to incorporate emerging risk, or having a designated point in the design at which it can be postponed due to an emerging risk that overwhelms it, is good practice.

In 2002, the then US Secretary of Defense, Donald Rumsfeld, in response to a question about weapons of mass destruction in Iraq, used the expressions "known knowns," "known unknowns," and "unknown unknowns." Although seemingly rather oxymoronic, the concept of "known unknowns" is well understood in risk management and can be traced back to 1955 (Luft & Ingham, 1955). The premise is that "known knowns" are those risks that are understood—that is, those that are observable or foreseeable. "Unknown unknowns" are those risks for which we have no understanding and are unable to predict on any level. "Known unknowns" are the most interesting in that they represent the risks that we know we have no current understanding of but, in all probability, have some ability to make a reasonable judgement on using past experience. In the case of the global pandemic of 2020, it was entirely foreseeable that a pandemic was probable, but the effect on the global economy was, perhaps, less well considered. Previous pandemics, such as SARS in 2003 and Ebola in 2013, did not cause the widespread economic meltdown that was witnessed in 2020 and, therefore, there may have been a certain complacency among nation states to implement appropriate risk management strategies. A case, perhaps, of global normalcy bias.

Identifying emerging risks, and understanding their potential effect on any given project, is important in being able to define appropriate design solutions. The effect of a global pandemic is unlikely to affect the way in which motor cars are designed, for example, but could have implications for a public building such as an airport or cruise terminal where new security protocols have to be introduced, or distances between people must be increased, or air conditioning systems have to be enhanced. Conversely, the motor industry has been affected in recent years, after the introduction of keyless entry systems, by technology used by criminals to clone an owner's key from a distance and steal their car. If one thing technology has taught us over the years is that whatever security measure one person invents there will always follow some form of counter-measure. This is a fundamental of the human condition and should perhaps be more widely understood by designers everywhere.

Another label attributed to risk, and specifically the management of risk, is the "black swan" event we mentioned earlier. This term, originally coined by Nassim Nicholas Taleb in his book, *The Black Swan* (Taleb, 2007), refers to events that are unpredictable in their nature and yet hugely consequential in their outcome. Taleb cites examples such as the meteoric rise of Google and the attack on the UN Trade Centre in New York in September 2001. The unpredictability of the event and the size of its impact are two of the three traits that Taleb asserts are evidence of a true black swan event. The third is that, post-event, we concoct explanations afterwards that make the event seem less random than we initially observe it to be. This has much to do with a type of behaviour known as hindsight bias.

Hindsight bias is the seemingly rational explanation for something that we feel we had already foreseen, summarized best perhaps by saying "I knew that all along," or "I told you so." An example perhaps of this is the accident at the Fukushima Daiichi nuclear power plant in Japan in March 2011. After a magnitude 9.0 earthquake at sea, some 175 km from the plant, a 14-m-high tsunami breached the sea defences of the plant which had been designed to only resist water to a height of 5.7 m. The in-rushing seawater overwhelmed electrical backup generators that were providing emergency power to the pumps cooling the reactors. The reactors had shut down automatically after the initial earthquake as part of the plant's emergency response protocol. Without this essential cooling, the reactors quickly overheated.

In a subsequent interview, Mr. Naomi Hirose, president of the Tokyo Electric Power Company, which operated the Fukushima plant, admitted that "measures could have been adopted in advance that might have mitigated the impact of the disaster." Such measures could have included fitting waterproof seals on all the doors in the reactor building, or placing an electricity-generating turbine on the facility's roof, where the water might not have reached it. Considering the level of planning, design, and safety analysis that is inherent in the building of nuclear plants, these measures seem, with hindsight, relatively simple and straightforward. Japan is a country that is historically given to suffer from seismic events and when these events occur at sea, they cause tsu-namis (the word "tsunami" is, after all, Japanese and means "harbour wave"). We are also aware that climatic events have become more severe in recent years, a trend that appears set to continue; and large, complex infrastructure is designed for a lifespan of several decades, meaning it may have to be over-designed in today's terms just to con-tend with possible conditions in the decades to come.

A few examples of other designs that failed due to perhaps not wholly appreciating, or adequately mitigating for, the operational environment are as follows:

- Tacoma Narrows Bridge, 1940: collapsed; suffered from inappropriate aerodynamic design in a sustained, moderate wind.
- Port of Ramsgate walkway, 1984: collapsed; not adequately designed for a marine environment.
- Millennium Bridge, 2000: re-engineered; insufficiently designed to prevent lateral motion caused by people walking across the bridge.

It should be noted that although these explanations have been greatly simplified for the sake of brevity, the overarching failure of each one stemmed from inadequate or inappropriate design considerations.

It is perhaps poignant to return to the words of Mr. Naomi Hirose. His observations on how cumulative measures can possibly provide better mitigation than large, single measures reveal the core of the designer's role in creating products that are safe to operate in all conditions.

> "*Try to examine all the possibilities, no matter how small they are, and don't think any single counter-measure is fool proof. Think about all different kinds of small counter-measures, not just one big solution. There's not one single answer.*"

To ensure that adequate regard has been paid to the external influences that may affect any project, the following could be considered as a minimum.

1. Due regard should be paid to any previous product either in use by the client or available in the general market, specifically to any previous design parameters that created safety, operational, or maintenance issues.
2. The advice of specialists should be sought where doubt remains in such areas as regulation, international standards, marketing, and so forth.
3. Manufacturers and suppliers of materials and equipment due to be commissioned into the project should be engaged to discuss possible and practical alternatives.
4. A thorough design risk assessment should be completed that examines the four principal stages of the product life cycle—design, production, use/maintenance, and disposal.
5. The client should record all relevant risk to the project in a risk register that should be updated as necessary.
6. Decisions on risk control measures should be recorded in the project's document package.

Control Measures

In order to mitigate the likelihood of a risk being realized, it can be necessary to provide an array of mitigation types in order to ensure that the risk is prevented so far as is reasonably practicable. The "Swiss cheese" model (Reason, Human error: models and management, 2000) graphically demonstrates this as a series of layers of Swiss cheese with distinct holes in them (see Figure 4.7).

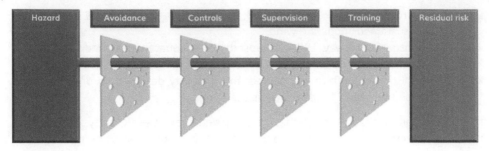

Figure 4.7   James Reason's Swiss Cheese Model.

Each layer (or slice of cheese) represents a form of mitigation or control measure that prevents risk from causing harm. The holes in the slices represent events where the particular control measure that the slice represents fails for whatever reason: a person makes an error of judgment; a training course is not completed; a machine guard fails, and so forth. Ideally, where risk "breaks through" any particular "hole" in the control measure there would be a further measure beyond it which prevents the risk from escalating. Occasionally, however, Lady Luck (or more likely, bad planning) conspire to arrange the holes in the slices so that they line up, allowing the risk to bypass all the control measures and thus cause harm in some way. This is why several "layers" of smaller, efficient control measures can often be more effective than a single, complex one. The types of control measures that can be applied are discussed in the general principles of prevention section of this chapter.

Risk Identification Tools

Although perhaps very few designers may find themselves in a position to officiate over such a project as one involving nuclear power, it is a fitting reminder that we should, in all cases, be considering the "what-if" questions. In engineering disciplines there are a number of tools that can be used to answer those what-if questions, often to a very high accuracy. Up to now, we have considered risk in general terms of likelihood and severity: this is known as *qualitative* risk assessment as it only considers the presence of risk. For some designs there may be a need to know calculated levels of risk, or to perform a *quantitative* risk assessment. This may be due to the need to address the mean-time-between-failure (MTBF) rates of certain components, or accurately predict the likelihood of some event occurring. There are a number of methods for doing precisely that which engineers in many different disciplines use. They include:

- failure modes and effects (and criticality) analysis;
- fault tree analysis;
- event tree analysis;
- probabilistic safety analysis;
- the "Bow Tie" method.

Failure Modes Effects (and Criticality) Analysis

Failure mode effects analysis (FMEA) and failure mode effects and criticality analysis (FMECA) are used to identify the ways in which a system could potentially fail (BSI, 2018). FMECA is essentially an extension of the FMEA method where levels of failure are calculated in order to prioritize and manage the resultant effects. These analysis tools are applied in two stages. Firstly, all the ways a system could fail are identified, along with the potential effects that each failure might cause. Secondly, these failures are critically analysed in order to establish a ranking of the probability and severity associated with each.

The information gained through these analysis tools can be used to inform the design management process so that the higher priorities are investigated first. It is a useful tool to understand failures associated with a particular system, how that system sits within other more complex systems, and how each area affects the higher-level systems.

Fault Tree Analysis

Fault tree analysis (FTA) is a failure-orientated method used either quantitatively, to show the probability of a top-level event occurring, or qualitatively to indicate what must happen in order for the top-level event to occur. FTA helps to establish the events *leading up to* a substantive failure of the system.

The FTA process begins with the definition of the system boundaries; this will ensure that the focus remains within a particular area. This is followed by the identification of a specific high-level failure, also known as a top-level event. The next stage is to consider all the failures (low-level events) and combinations of failures (sub-events) that could lead to this high-level failure occurring.

The tree of faults is held together with "AND" and "OR" logic gates. With "OR" gates signifying only one of the inputs needs to occur and "AND" gates meaning all of the inputs must happen before the sub-event occurs.

An example of a qualitative FTA might be what could lead to a machine overheating (see Figure 4.8). Whereby the possible failures (low-level events) could be "oil feed sensor fault," "oil feed pump fault," and "electrical fault." To cause the machine to overheat, either the bearing must overheat or there must be an electrical fault. To cause the bearing to overheat there needs to be both an oil feed sensor fault and an oil feed pump fault.

A qualitative FTA is used to identify all the possible events and combinations of events that would result, if not mitigated, in the high-level failure. Whilst a quantitative FTA follows the same construction, it also includes the probability of each event occurring. This, with the combination of events through "AND" or "OR" gates, provides a mathematical means of calculating the likelihood of the top-level event occurring.

Although FTA is a common engineering methodology used to identify the likelihood and dependencies in potential system failures, it still requires root-cause analysis tools and methods to help in the identification of cause and effects, such as FMEA and FMECA amongst others.

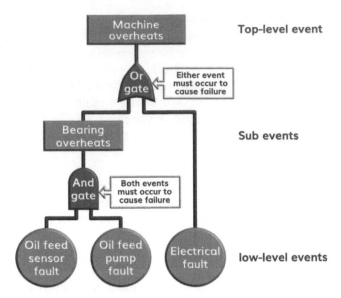

Figure 4.8 Fault Tree Analysis Diagram.

Event Tree Analysis

Event tree analysis (ETA) is a process that begins with a top-level event (possibly identified through having conducted a fault tree analysis) and goes on to identify a range of possible resultant outcomes, whilst also taking into account any potential mitigating actions that have been put into place. Each mitigation has a probability value of between 0 (where it is 0% effective) and 1 (where it is 100% effective). Any value in-between these identifies the level of *possibility* of the outcome occurring. It is therefore possible to calculate the levels of probability for each sub-event occurring following a top-level event. The calculation of probabilities should be undertaken by qualified practitioners or provided by the product manufacturer. The probabilities given in the following example are purely hypothetical for ease of explanation. ETA helps to establish the events that might occur *after* a substantive failure of the system.

Using the top-level event "machine overheats" (from the FTA example), the probability of this causing a fire to start is established as **0.5** (50%) (which is to say that there is an even chance that the overheating machine will cause a fire), whilst the probability of the automatic spray system putting the fire out is established as **0.8** (80%) (or reasonably likely that if operated it would put out a fire). So, if there is a **0.5** probability of the fire starting and a **0.8** probability of the fire being extinguished, then the overall probability of the overheating machine causing a fire that is then extinguished is **0.8 × 0.5 = 0.4** (or 40%). Therefore, the probability that a fire starts, is not put out by the spray system, and continues to grow is **0.2 × 0.5 = 0.1** (or 10%) (see Figure 4.9).

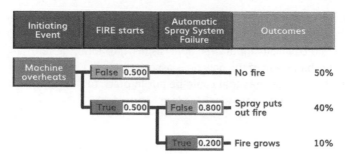

Figure 4.9 Event Tree Analysis Diagram.

Probabilistic Risk Assessment

Probabilistic risk assessment (PRA), also known as probabilistic safety analysis (PSA), is a logical and all-inclusive methodology used to evaluate all the risks that are associated with complex systems. The risk assessment is based on three main areas: what could go wrong; how severe it could be; and how likely this could be. Common methods to quantitatively determine the answer to the last question are FTA and ETA. PRA/PSA are associated with nuclear, chemical, aerospace, and other highly complex industries. The calculation of risk probability is used to assess the levels of safety associated with a system and to prioritize mitigation effort (Springer Series in Reliability Engineering, 2010).

Bow Tie Method

The Bow Tie method visually represents the threats and consequences associated with known hazards and is in effect a combination of the two analysis tools described previously (see Figure 4.10).

The methodology starts with the identification of a hazard (something that has the potential to cause harm) and the definition of a "top event" (i.e., what could realistically go wrong).

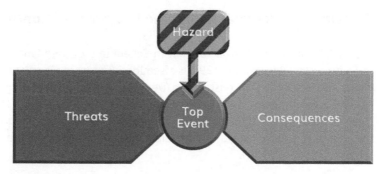

Figure 4.10 Bow Tie Diagram.

The next stage is to look at all the threats that could lead to the top event and what, if any, preventative measures could be put in place to reduce the threat. These threats could be taken directly from a fault tree analysis (FTA) to form the left-hand side of the bow tie. The right-hand side of the bow tie is associated with the possible consequences and any preventative measures that could be put in place to reduce them and could be deduced from an event tree analysis (ETA) (see Figure 4.11).

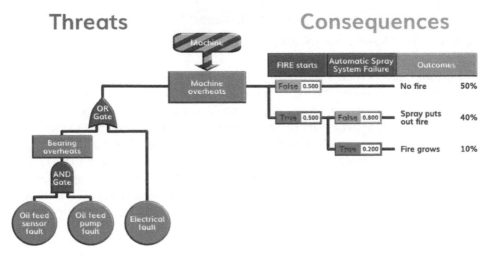

Figure 4.11   Bow Tie With FTA and ETA.

Whilst all of these and other methodologies may be useful in providing *predictive* outcomes, it is still incumbent on the designer—or design engineer—to, firstly, engage the correct methodology to give a valuable prediction and, secondly, to *ask the right questions*.

General Principles of Prevention and the Hierarchy of Control

The general principles of prevention, as laid out in Schedule 1 of the Management of Health and Safety at Work Regulations 1999, is a sequential process for risk mitigation. These principles also feature in the Construction (Design and Management) Regulations 2015 in Appendix 1 and stem originally from Article 6 of the European Union Directive 89/391/EEC to encourage improvements in the safety and health of persons at work throughout EU member states. The general principles of prevention are demonstrated here as a cascade (see Figure 4.12).

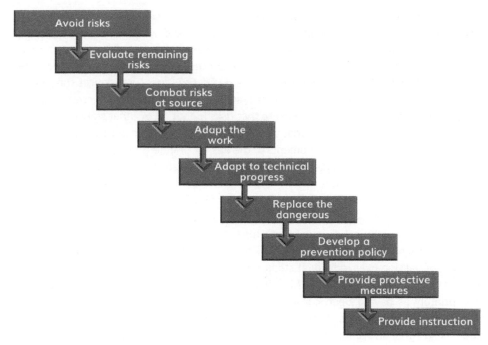

Figure 4.12 General Principles of Prevention.

The general principles of prevention exist to reduce the risks associated with any number of hazards and although they originate from a European Union directive concerned with the health, safety, and welfare of people at work they can be readily applied to any risk management strategy. In the regulatory environment chapter, we discussed, as part of the section on the Health and Safety at Work etc. Act 1974, the meaning of the phrase "so far as reasonably practicable" and its implications in health and safety law. Another phrase used interchangeably with this one is "as low as reasonably practicable," particularly when we are dealing with the general reduction of risk, and it is this phrase—and particularly its abbreviation, "ALARP"—which is more commonly used. It is important to note the use of the word "practicable" as opposed to, say, "practical" or "possible." The general principles of prevention are intended to provide a pathway to reducing risk to a level as low as reasonably practicable but, it should be noted, that wherever there is a hazard, there will always be a risk. To recap, a hazard is something with the potential to cause harm and risk is the quantum of likelihood and severity that stems from it. A hazard can be a physical thing, such as electricity, or it can be less tangible such as financial speculation.

In the guidance to the Construction (Design and Management) Regulations 2015—L153 (HSE, 2015) these principles are summed up as "eliminate-reduce-control" which, although considerably simpler, still encompass the *spirit* of the principles of prevention if not their exact list of precedence.

Each control measure should be considered in a contiguous sequence, from first to last. This is a flow of control from beginning to end which should, when properly considered, reduce the risk in the final product to as low as reasonably practicable. Where any particular step cannot be addressed within the design process then the next step should be considered. It should be noted that even where one control measure reduces the risk, in the eye of the designer, to a level as low as reasonably practicable, any successive steps should still be considered. This is to ensure that risks identified by other stakeholders are addressed. It should also be noted that the process of following the general principles of prevention is a cyclical one—amendments to the design, or emerging risks, or risks identified late in the process by additional stakeholders will all cause the process to "reset" and begin again.

The process of following the general principles of prevention should also be modified to fit the particular size and complexity of the project. A project that is encompassing the very latest in technical developments, for example, is unlikely to benefit from extra time being spent on adapting to technical progress. The designer should be cautious, however, of passing over individual steps simply because they appear at first glance to be irrelevant to the project and this is where it is important that the steps are recorded appropriately as having been considered. This is one of the functions of the design reviews, held at the various stages of the design process and where all relevant stakeholders can agree on the integrity of the design as it matures. Additionally, the proactive use of a risk register can help to not only identify, but also provide control measures for, emerging risks that arise from decisions made during the design process. Control measures, and the considerations that led to them, can also be recorded in a design risk assessment for the project.

The National Institute for Occupational Safety and Health in the United States of America identifies an alternative group set for control measures, and their hierarchy is the following:

- Elimination—physically remove the hazard.
- Substitution—replace the hazard with something less hazardous.
- Engineering controls—isolate the hazard through mechanical means.
- Administrative controls—change the way people work, or train them sufficiently.
- PPE—protect the worker with personal protective equipment.

Although slightly different to the general principles of prevention, this hierarchy is similar in its determination to reduce risk, albeit in a less prescriptive way. It is often used in health and safety circles where the resultant abbreviation—ESEAP—is perhaps more readily remembered than that for the general principles of prevention. A representation of this hierarchy demonstrates the effectiveness of each step in controlling risk (see Figure 4.13).

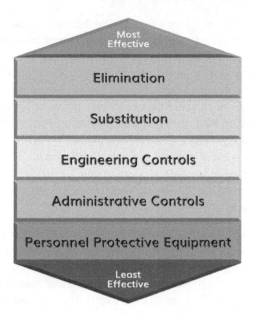

Figure 4.13 Control Measure.

The relative effectiveness of the control measures in the general principles of prevention is the same: the first, "avoid risks," is clearly the most effective because if the risk is removed there should be no harm possible, whilst the last, "give appropriate instructions," is the least effective as it is the most easily ignored or adulterated. This methodology of implementing control measures is seldom used anywhere else except by health and safety professionals due to it being a requirement of health and safety legislation. It should be reasonably supposed, however, that any designed output that is to be used in the workplace, or interacts in any way with people at work—be they cleaners, operators, engineers, and so forth—will be governed by this type of legislation at some stage. For this reason alone, it is beyond prudent to ensure that the design process, from the concept stage through to the final delivery, encompasses the provisions of that legislation so far as reasonably practicable. Other prudential reasons for following the general principles of prevention during the design process are as follows:

- The health and safety of all persons interacting with the product (operators, engineers, cleaners, maintenance personnel, etc.) should have been considered.
- The number of control measures needed to be put in place by the end user will be minimized.

- The supervision of safety by the end user will be based on robust control measures.
- The amount of administrative intervention (supervisors, documents, training, etc.) by the end user will be minimized.
- The documentation for operation, maintenance, and cleaning procedures will be more favourable due to the minimization of risks inherent in the product and this should therefore have a lower financial impact during its serviceable life.
- The safety of those who have to modify, dismantle, or dispose of the product at the end of its serviceable life will have been considered.
- Adaptations to the product during its serviceable life should have inherently less risk and therefore should be achievable at a lower cost.
- The need for personal protective equipment should be minimized with a resultant lower financial impact.

Our responsibility to provide design solutions that are as risk-free as possible—or, so far as reasonably practicable—is, as we have seen, driven by regulatory, moral, and ethical reasons. Legislation regarding the safety and welfare of people has been in existence for a long time, with arguably the earliest statute that could be said to be "health and safety" law being the Health and Morals of Apprentices Act 1802. This legislation, however, was reactive to the appalling working conditions that children were exposed to in the textile factories of the time. The proactive legislation we have today is more focused on the elimination or reduction of risk at source, hence the role of the designer in designing-out risk.

Further reinforcement of the design process being a fundamental source of risk prevention can be seen in the years since the Health and Safety at Work etc. Act 1974 through such things as the standards we have discussed—for example, BS 7000—as well as initiatives like the Get It Right Initiative. All of these combined create a governing framework within which we must provide the safest design solutions for any given project.

It is improbable, however, that we can eliminate *all* risk at the design stage. Risk will always exist where there is a hazard. If we design a machine that is electrically powered, we have the hazard of electricity to consider when considering risk control measures. If we were to design the same machine without a power source—that is, it is to be manually operated by a person—we still have the hazard of the machine itself. The risks in this case might be musculoskeletal strain caused by operating the cranking handle, or the physical hazard of being crushed or drawn-into the machine's gears or conveyor. Risk mitigation—using the "eliminate, reduce, control" mantra—should therefore be a sequential process that is reset wherever there is a functional change in the design, or even where a control measure is introduced. Only by mitigating each step in this way, causing the risk to become ever more tightly controlled, can we adequately say that the risk is reduced to as low as reasonably practicable (see Figure 4.14).

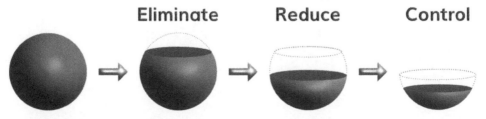

Eliminate **Reduce** **Control**

Figure 4.14 Hierarchy of Controls.

Let us consider the application of the general principles of prevention in terms of design to understand how they may be implemented (see Table 4.2).

Table 4.2   Principles of Prevention—Possible Applications.

Principle of prevention	Possible application
Avoid risks	Clearly, the best way to avoid risk is to remove it completely. We understand, however, that this is not possible completely. Racing cars, for example, travel faster today than ever before and yet the risk of drivers being injured or killed is greatly reduced due to the creation of strong "cells" in which they sit. This has been possible due to the creation of advanced materials such as carbon fibre and the design of crumple zones, which progressively deform in the event of an impact to reduce the forces of deceleration. The risk of collision, however, remains.
Evaluate risks that cannot be avoided	Risks may be described as observable, foreseeable, and emerging. Emerging risks are often the hardest to mitigate as they often require a certain amount of speculation. A risk register can help enormously in the identification and monitoring of risk throughout a project. Any organization should already have a corporate risk register and often this can augment a project's own risk register. For example, an organization selling products abroad may be aware that regulations in a foreign country are being changed by that nation's government, which may impact on the design of products or services for export. The European Union's General Data Protection Regulation of 2018 is an example of this.
Combat risks at source	In combating risks at source, we can deploy a hierarchy of control based on eliminate-reduce-control. To eliminate persons from risk, a process could be automated instead. To reduce the risk further, the software running the automation could remotely call-in exceptions to the manufacturer so that an engineer can be sent to repair it. To control the risk further still, the enclosure to the automation process could be fitted with locks to which only the manufacturer's engineers have the key.
Adapt the work to the individual	Ergonomics and anthropometry play a vital role in the design of machines and processes where workers are required to interact with them. Relying solely on science, however, may not provide the whole picture. Perhaps by engaging with the human resources or health and safety professionals of an organization further insight could be gained of the actual risks of operation by examining sickness and accident records.
Adapt to technical progress	Motor cars up until fairly recently were designed to be powered by internal combustion engines. Today, more and more, each new model of car has been designed to be powered by alternative power sources and the allowance for the space required for them has had to be designed in from the outset. Similarly, we may design a building with a stronger roof than necessary to allow for the fitment of photovoltaic cells (solar panels) in the future.
Replace the dangerous	Dangerous materials or processes are not confined to those affecting individuals. The industrial chemical trichloroethane was produced in vast quantities before it was discovered to have serious ozone-depleting effects. It has been replaced with more environmentally-friendly substances, such as *n*-propyl bromide solutions. Dangerous of course means the risk is "imminent" and therefore further measures should be installed to lessen that risk—for example, bunding for chemicals, guarding for machinery, lockouts for access points, and so forth.

(Continued)

Table 4.2 (Continued)

Principle of prevention	Possible application
Develop a prevention policy	When designing a product, there may be a need to produce maintenance procedures to ensure its ongoing safety and it is important to be cognizant of the skill set of the individuals who will undertake it and, therefore, the level of risk that this can in itself create. Aircraft engineers, for example, are highly trained and work to very exacting standards in pristine conditions: they are able to make precision adjustments when required. A machine for the public market might, alternatively, be better suited to having, say, sealed bearings instead of maintained ones in order to prevent them being overlooked by the operator.
Provide protective measures	Collective protective measures should always be considered over individual ones. In the design for a rooftop garden, guardrails with infills would be deployed where access is common whereas man-safe systems or harness anchor points could be used for infrequent access by trained maintenance engineers, for example. In any design, it should be the ambition of the designer to prevent the need for the operator to have to use personal protective equipment as this is the least controllable, and often least effective control measure.
Provide instruction	An instruction manual for the operator should consider not only the type of person that operator is likely to be (aircraft pilot, field engineer, member of the public) but also their manager, if any, and the organization for whom they work, if at all. Consideration of the style of language and complexity is important. Similarly, consider if operating procedures can be taken back to the design level. Instead of, for example, instructing the operator to wait for a pressure dial to reach a certain value before pressing the next button, consider installing a sounder that audibly informs the operator that the correct pressure has been reached.

CDM Deliverables in Support of Risk Management

The Construction (Design and Management) Regulations 2015 (CDM) are principally concerned with reducing the risk to the health and safety of people involved in the construction of a product as well as those who will be involved in its use, maintenance, and disposal. This is done through considering risks at the design, construction, and in-service stages and by designing-out the risk, implementing appropriate control measures, and/or introducing management controls. The identification and control of risk in CDM is done through effective management of information and documentation, some of which is mandatory.

As we have seen in the chapter on the regulatory environment, CDM covers far more than what we might consider "traditional construction," for example, building houses and other structures. If we extrapolate the requirements of CDM across all products that require a design and management function, we find a valuable system of design management that is effective in reducing error, wastage, and unnecessary cost. By simply altering the terminology, and applying the spirit of the regulations as well as the letter of them, we can readily see their relevance to all design projects. To summarize, the mandatory documents required by CDM can help to control risk by capturing and portraying relevant information throughout the project, ensuring that risk is dealt with appropriately at each stage (see Figure 4.15). The documents are:

- pre-construction information;
- construction phase plan;
- health and safety file.

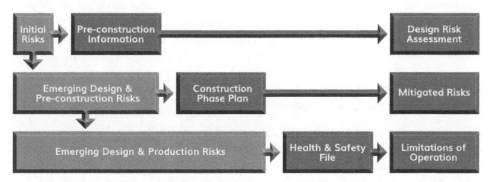

Figure 4.15 Project Risk Management.

These documents work in conjunction to mitigate risks from before construction commences to a point after the product has entered service. By sequentially identifying and documenting the risks, it is possible to introduce control measures at each stage and record them, thus enabling the control measures to be disseminated to those persons who will be affected by them. In the case of a building these might be surveyors, site workers, site visitors, users and maintainers of the building, and so on. Each stage deals with the risks identified in the preceding document, as well as identifying new ones in its own stage. Additionally, as CDM is essentially based around the fundamental requirements of UK health and safety law, the outputs at each stage are readily translated into relevant mandatory workplace safety considerations.

Pre-construction Information

The initial observable risks associated with the design and the intended physical environment in which it will be produced are identified and recorded. For a construction project such as a building, this might include information on:

- the organization of the project, key dates, and personnel;
- the site set-up, access and egress, and any potential security implications (nearby schools, roads, etc.);
- any adjoining works that may conflict with the project;
- any specific hazards such as asbestos, hazardous materials, or confined spaces;
- the location of utilities and any buried services;
- any environmental implications, such as the presence of protected species, or the need to protect against dust, noise, or other emissions.

These observable risks are intended to provide information for both the designer and the contractor undertaking the works. The designer can use the information in order to incorporate any appropriate mitigation into the design: the presence of nearby power lines might, for example, cause reflection on how tall the structure can be. The contractor can use the information to incorporate any appropriate mitigation during the construction stage: the presence of nearby fragile buildings might, for example, cause them to use compression piling instead of percussion piling.

The purpose, therefore, of the pre-construction information is to identify existing, observable risks that can be used to inform both the design and construction (or production) stages, thereby reducing potential costly issues from arising. It is this comprehensive identification of risk that can be translated across to any design project. We can think of the pre-construction information as the completion of the first part of a risk register—identifying relevant risks in order that suitable control measures can be determined and implemented. For all types of design project, the sort of information we could gather might include the following:

- The organization of the project, key dates, and personnel.
- Key objectives of the design, perhaps with reference to function, emissions, operability, aesthetics, and so forth.
- Any information pertinent to the design, perhaps from tests, surveys, or feedback.
- Any external influences on the project, in terms of regulatory, marketplace, supply chain, and so forth.
- The presence or potential inclusion of hazardous materials or techniques.
- The capability and capacity of the intended production facility or technique.

Construction Phase Plan

This document, as its name suggests, is the plan for how the construction stage will be conducted and is informed by the pre-construction information preceding it. It details how the salient risks identified will be managed or controlled, as in our example of changing the method of pile driving. As well as the risks identified in the pre-construction information, the construction phase plan also identifies, and proposes mitigation for, a number of risks that can be found on any building project, such as:

- the roles and function of managers and supervisors on site;
- the type, frequency, and location of site meetings;
- the arrangements for providing welfare facilities (for eating, resting, washing, and toilets);
- the arrangements for emergencies such as first aid, fire, evacuation, and so forth;
- the training and informing of on-site personnel and visitors.

By detailing all of these arrangements and control measures, the construction phase plan can become a necessarily substantial document. It can be thought of as the further completion of a risk register, where already-identified risks are provided with forms of mitigation and where further emerging risks are identified and recorded. In a sense the plan is transitionary, as it deals with both observable and foreseeable risks. The proximity of fragile buildings is an observable risk; that of poorly informed personnel working on site is foreseeable. It may also be likened to a safety case, where the product can be tested against known risks, emerging risks, and also the capability and capacity of the producer to actually produce the product to the accordant specification and tolerances.

When considering all types of design project, the sort of information that might be included in this document could be:

- the identity, roles, and responsibilities of key and additional stakeholders in the project;
- the type, frequency, location, and expectations of project meetings;
- the structure and designated format for the transfer of information and documentation;
- any external influences on the project, in terms of regulatory, marketplace, supply chain, and so forth;
- any methods and expectations for the way the product should be validated (confirmed fit for purpose) prior to entering use.

Health and Safety File

Being peculiar to CDM, this documentary requirement is notable for its primary concern for the health, safety, and welfare of people who interact with the product during its in-service stage. Its purpose is to inform the user of any residual risks that remain after the design and construction stages: that is to say, those risks that could not be eliminated or otherwise controlled. An example might be the presence of pre- or post-stressed concrete elements within the structure. A typical file for a building might include information on:

- description of the works;
- fire risk assessment and evacuation strategy;
- design risk assessments;
- test and validation certificates: for such things as electrics, gas, ventilation, water, data, and so forth;
- as-built drawings and calculations;
- cleaning and maintenance manuals, instructions, and schedules;
- safety information on structural elements, installed equipment, hazardous materials, services, and so forth.

Compiling the mainly foreseeable risks in this way allows the owner or operator of the building to create their own risk assessment using robust information. This greatly assists the efficacy of their own safe system of work. It can also be seen that this file forms the greater part of a risk assessment that could be of benefit during disposal, in that it identifies the location of hazardous materials and stored energy systems that could cause serious harm to those involved in the dismantling or demolition of the structure. The content of the health and safety file is notably similar to that required by the CE marking regime, commonly known as a technical file. This is not surprising as they are both concerned with demonstrating to the user that the product is safe to use, although a technical file's output to the end user is the issuance of a declaration of conformity rather than the detailed information that led to it. For all types of design project, this final documentation could include:

- description of the product;
- drawings of the product, including any relevant specific drawings, such as circuit diagrams;
- copies of test results or assessments and the validation of any critical components;
- instructions or information about the safe use of the product.

Competently Dealing with Risk

Engineering or design safety is prevalent, as one would expect, in the concept and design stages of a product (see Figure 4.16). During production of the product both engineering safety and health and safety tend to have equal input. Traditionally, however, health and safety is involved at this stage due to the actual production (or construction) activity itself, such as: in the assessment of risk; the creation of safe work processes; and the observation of compliance with legislation, and so forth. Engineering safety during this phase is, conversely, concerned with the actuality of constructing the product and adapting the design to emergent issues with it. Generally, only health and safety remains during the operational life of the product.

Figure 4.16 Silo Safety.

One common failing during the design process is not involving health and safety practitioners from the outset; instead relying solely on engineering safety specialists to advise the designers. Whilst some engineering safety specialists may have experience of operating and maintaining products in the field, their careers may have kept them away from the "factory floor." This leads to situations where those responsible for ensuring that the design is safe to operate may have had limited operational experience or understanding of the operators' needs and experiences. Moreover, those that ensure that the product is operated safely, that is to say, health and safety practitioners, have little or no input on how effecting even a minor design change could help to reduce the operational risks that operatives face on a daily basis. This cause-and-effect dilemma is a result of "silo" thinking among departments or divisions within an organization, which may be entirely natural behaviour where humans are grouped together, but it is behaviour that stifles true design safety and should be countered. This isolationist approach to safety also ultimately incurs greater risk to those involved in the operation and maintenance of the product as they may well be reliant on processes and personal protective equipment to prevent harm instead of engineering measures, contrary to the hierarchy of controls we have discussed.

This demonstrates how the early stages of design need input from health and safety practitioners to provide advice on the type and range of issues that operators are faced with, as well as any lessons learned from similar products in the past. They can also provide information on safety procedures that have had to be implemented, or accident data that highlight where failures in design might have caused loss or harm to occur in the past. They should also be able to advise on any applicable legislation or forthcoming updates or amendments.

Similarly, engaging with engineering safety specialists on an existing product, prior to the design of an updated version, will allow them to see where potential design shortfalls have occurred and how these might be mitigated in future designs. Often, simple design changes can yield robust, long-term risk reduction and perhaps prevent the need for specialist or protective equipment whilst in use. This is also highly valuable in regard to maintenance procedures where products, particularly machinery, are run in an "exception" state, where designed-in safety features may need to be disabled in order to either maintain it or to trace and rectify problems.

This approach calls for a more gradual interaction between both engineering and health and safety disciplines (see Figure 4.17) in contrast to the "step-in, step-out" approach that is all too familiar, and which can be deleterious to the design process overall.

Figure 4.17 Integrated Safety.

This approach clearly requires not only the right information being available to the right people at the right time, it also requires those people to have the relevant skills, knowledge, and experience too. That is to say, to have the competency to carry out their function properly and robustly. Added to which, the commitment to safety must be a shared and common goal of all stakeholders in the project. There is little point in the engineering safety and health and safety experts discussing solutions if they are ignored.

The competency of individuals and the presentation of safety information for designs has been greatly influenced in recent years by two very important reviews, both as a result of major catastrophes. The first, the Nimrod Review (Haddon-Cave, 2009), was the result of the RAF Nimrod MR2 Aircraft XV230 that crashed in 2006, and the second was the Building a Safer Future—Proposals for reform of the building safety regulatory system report (Ministry of Housing, Communities and Local Government, 2018), which was conducted as a result of the Grenfell Tower fire.

Even though the reports covered two disparate design types, one for a complex military aircraft and the other for a high-rise building, there are a number of similarities in their findings. Both reports identified issues associated with the competency of individuals and both identified the issues surrounding the collection, management, and presentation of safety-critical information, and both identified the need to ensure that safety cases are created and managed by competent persons throughout the lifetime of the product. Despite the differences between the two products, the conclusions were that the appropriate and timely collection of safety information, started at the design stage, should be properly managed and made available throughout the product's life until final disposal.

This information can be contained in a safety case, the choice of which is based on product type and which can require specialist input in creating. Much of this information though is also pertinent to the health and safety file, which can be readily created for *any* type or design project. This information is available throughout the design process and requires only simple marshalling and management in order to create a suitable document or repository. Again, the four Cs are the guiding principle for this: adequate *control* of the information at hand; a *competent* person to manage it; the *cooperation* of all stakeholders in the project to provide relevant information and; a robust system of *communication*, both throughout the project and during the entire lifetime of the product.

The creation, management, and dissemination of this information is also in line with the requirements of BIM and of the "golden thread" for high-rise buildings (over 18 metres or 6 floors, as well as possibly for buildings that house vulnerable people) (Ministry of Housing, Communities and Local Government, 2018).

Risk Management Summary

There are a number of influences on the risks associated with the design of a product up to the point at which it is accepted into service. Any that cannot be mitigated will result in residual risks that will impose either limitations of operation or the need for operational risk assessments. These residual risks must be recorded and formally passed to the client at product acceptance. General influences on design risk include:

- level of risk associated with the initial design option;
- issues identified during the concept design process;
- issues identified as the design matures;
- issues identified with the production stage;
- issues identified with the potential operation of the product (possibly by a wider group of stakeholders during later design reviews).

Any of these issues could be realized in various categories such as political, operational, environmental, regulatory, and so forth.

Mitigation can be achieved throughout the design life cycle by adopting the general principles of prevention or a suitable alternative hierarchy of control. It should be remembered that the process of controlling risks is not a single step though; rather, it is a continual process throughout the design process, and can be greatly enhanced by conducting design reviews at the end of each design phase to ensure that the cycle is complete and that all risks have been identified, as far as is possible. This can be crucial especially where complex projects require several interconnected design solutions to be created. As each design solution is proposed, it is important to document its effect on other solutions to ensure that conflictions or additional risks are not being introduced.

Risks that cannot be fully mitigated during the design stage prior to acceptance into operational service will most likely impose limitations on the use on the product. A motor car, for example, may have any number of control measures due to the risks from being driven by a human operator, hence the reason for seat-belts, anti-lock brakes, crumple zones, and so forth. The limitations of use, usually documented in the operational manual (not being used off-road or overloading the roof rails, for example), are there to further control risks to a tolerable level within the design scope of the product. This level is normally associated with the term "as low as reasonably practicable."

As we have discussed, *as low as reasonably practicable* must be demonstrable through cost-benefit analysis to show that any further mitigation would be so costly as to be grossly disproportionate to the level of reduction in risk. This is the balance between the quantum of risk and the additional cost, time, and effort required. It is important that the risks throughout the operational lifetime of the product are managed and this begins with the design process. It is continued by the product's owner/operator as mitigations become cheaper or alternative options become available. For example, the general principles of prevention include adapting to technological advances and an obvious instance of this is the technology developed in Formula One and other motor sports, which has, over the years, been adopted by mass-production car manufacturers to the benefit of the safety of all road users.

For risk management to be effective, it should consider the following:

1. The identification of all risks—both observable and foreseeable—to the design.
2. Consideration of possible emerging risks (or "black swan" events) that may overtake the design or the project. It may only be possible to discuss and document this consideration as control measures may be hypothetical at best.
3. An assessment of those risks to consider likelihood and severity as well as who or what will be affected by any potential outcomes.
4. Clear documentation that identifies the owner of each risk.
5. Any control measures are disseminated to the relevant parties for inclusion in the design or other risk registers.

5

Effective Design Strategy

The Importance of an Effective Design Strategy

So far we have looked at the wide range of factors that can influence the design process and the many considerations that should be taken into account. We have discussed the importance of the designer's role in assembling sometimes quite a large amount of data in order for them to provide the most suitable design solution. We have seen that the relevance of this solution is important for a number of reasons: those of propriety; professionalism; ethics; cost benefit; operability and maintainability; and, above all, safety.

We have seen how the management of risk can be an effective tool in providing design solutions to any project and how it is important to identify and mitigate risks effectively. Equally important is how engaging a broad range of stakeholders in the project can result in much more successful design outcomes; ones which fulfil many more ambitions for the design than might otherwise be possible. The addition of information gained from previous "lessons learned" reviews can also shed valuable light on possible pitfalls and issues that may have hampered previous design processes that any of the stakeholders may have encountered.

We have seen that the concept of "safety" is one that we can apply to the many demands on the project other than those traditionally thought of as being safety related. And we have seen that safety, risk management, and good design outcomes are inextricably interrelated.

Moreover, we have seen that the designer's role in industry, society, and technology is not only critical but should also be highly prized. It is most important that we commit ourselves to designing products, irrespective of their use or intent, of the highest possible quality, suitability, and sustainability. And that we do so with the highest levels of professionalism and regard for our clients, suppliers, producers, and end users. This is true regardless of the size, type, and complexity of the design, the project, or the intended output, and the reader is encouraged to adapt all their experience, in combination with what has been discussed throughout this book, to each design project they are engaged on. The route from the initiating need for a product to its final disposal is shown in the form of a railway line, or route map (see Figure 5.1), for this is precisely what design is: a journey from an idea, through a productive, effective lifespan, to an eventual planned end. Although not stipulated for every design project, this "journey" provides a template for an effective design strategy. In this final chapter we shall take this journey in order to apply all that we have discussed so far.

An Effective Strategy for Safe Design in Engineering and Construction, First Edition.
David England & Dr Andy Painting.
© 2022 John Wiley & Sons Ltd. Published 2022 by John Wiley & Sons Ltd.

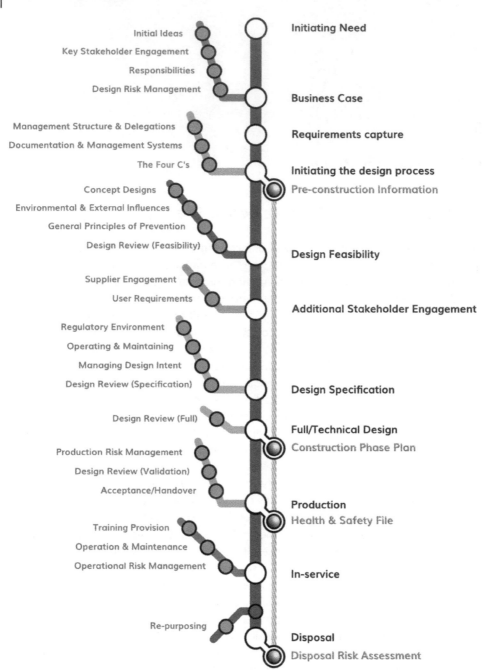

Initial Ideas
Key Stakeholder Engagement
Responsibilities
Design Risk Management

Management Structure & Delegations
Documentation & Management Systems
The Four C's

Concept Designs
Environmental & External Influences
General Principles of Prevention
Design Review (Feasibility)

Supplier Engagement
User Requirements

Regulatory Environment
Operating & Maintaining
Managing Design Intent
Design Review (Specification)

Design Review (Full)

Production Risk Management
Design Review (Validation)
Acceptance/Handover

Training Provision
Operation & Maintenance
Operational Risk Management

Re-purposing

Initiating Need

Business Case

Requirements capture

Initiating the design process
Pre-construction Information

Design Feasibility

Additional Stakeholder Engagement

Design Specification

Full/Technical Design
Construction Phase Plan

Production
Health & Safety File

In-service

Disposal
Disposal Risk Assessment

Figure 5.1   The Design Process.

Initiating Need

At the beginning of any project where there is a requirement for some element of it to be designed, there will be an initiating need. Some examples might be:

- the country needs to reduce its CO_2 emissions from the increase in power demand;
- the company needs more office space;
- the navy needs a more effective deterrent to modern threats;
- the organization needs to rebrand its home office equipment to remain competitive;
- the business needs a new range of vehicles to break into an emerging market.

Howsoever the need arises, and from whomever it originates, there will be drivers that underpin it. If there is an in-house design team, then these drivers could already be understood. If a designer has yet to be appointed, then they should appraise themselves of these drivers at the earliest opportunity. The driving forces behind an initiating need will be inextricably linked to the client's ambitions, perhaps in terms of growing the business, diversifying a product range, or capturing a new market. There are several examples of what these driving forces might be (see Table 5.1).

Table 5.1 Ambitions and Drivers.

Initiating need	Possible drivers
Reduction of CO_2	<ul><li>International treaty</li><li>Public fears and demands</li><li>Scientific evidence</li><li>Growth in population</li><li>Historic power outages</li><li>Reliance on foreign powers</li></ul>
More office space	<ul><li>Organic growth of the business</li><li>Dilapidation of the existing premises</li><li>Brand-conscious decision</li><li>Preparing the company for sale</li><li>Presence of hazardous material in the existing building</li><li>Plans to dynamically grow the business</li></ul>

(Continued)

Table 5.1 (Continued)

Initiating need	Possible drivers
Effective naval deterrent	• Changes in the global political climate • Emergence of new threats • Dilapidation of existing vessels • Catastrophic failures among existing vessels • Changes in political will of the government • Advances in defence technology
Rebrand of equipment	• Injection of capital from new owner • Injection of passion from new chief executive • Preparation of company for sale • Feedback from marketing and sales departments • Falling sales in core markets • Change in public tastes and trends
New range of vehicles	• Boardroom demands for improved sector share • Reliance on outdated technology • Regulatory changes in existing markets • Renewed public desire for the brand • Identification of sales potential • Sluggish sales in existing markets

These drivers will have an effect on the possible suitable design solutions proposed at the concept stage by helping to exclude those solutions that do not support them. For example:

- The reduction of CO_2 emissions from power generation can be readily achieved by the construction of solar panels. If the growth in demand is the driver, however, then this will be impractical during the hours of darkness.
- More office space could be achieved by increasing the existing building. This will not be suitable, however, if the existing building is structurally unsound.
- Repurposing existing naval vessels may not be practical if advances in weapons technology cannot be incorporated into them.
- An all-encompassing rebrand of equipment, including the engineering and technology within them, may not be achievable due to lack of funds. This may be true if the company is being primed for sale, and is to appear as financially healthy as possible to potential new owners.
- Wholly electrically-powered vehicles may not be suitable for an emerging market for which there are few charging points, or where the power infrastructure is incomplete or poorly maintained.

During this stage, many potential solutions should be considered in order to allow challenges to be made against the initiating need of the client. This is not to say that novel, bold, or unorthodox solutions will always be the solution, but it will allow flexibility in the design thinking that may, on occasion, be precisely what is needed. Understanding the drivers behind the initiating need will help to question these concepts thoroughly. But the concept stage is still a little way off yet, and we must now turn our attention to the next necessary stage of the project.

The initiating need should consider that:

- It includes a description at the highest possible level of what the client actually needs. ☑

- It is created with the express permission and inclusion of the key decision maker(s) for the organization or project. ☑

- It is expressed in a way that supports the client's strategy and goals. ☑

Business Case

Initial Ideas
Key Stakeholder Engagement
Responsibilities
Design Risk Management

Initiating Need

Business Case

We make business case decisions all the time, both in our private lives and in a work capacity. It is merely the process of balancing a potential outcome against potential risks—of weighing up known factors against unknown ones—and deciding on the best course of action that most suits a desired ambition. A business case is supported by conducting a cost-benefit analysis that evaluates the financial implications of a particular course of action. Spending, for example, £100,000 on a project might yield a return or a saving of £500,000, whereas doing nothing (zero cost) might yield a loss of £150,000. Or perhaps replacing an old, unreliable car with a new one might mean fewer breakdowns, but it will come at a higher cost. Whether that cost can be justified against the cost and inconvenience of sitting at the side of the road with the bonnet up is a simple expression of a business case. Corporate business cases might revolve around decisions about future expansion of production, opening up new markets, changing a product range, or preparing a business for sale.

The business case for a design project is no different. It is merely another way of evaluating an assessment of risk, except that we are dealing with the terms *cost* and *benefit* rather than *likelihood* and *severity*. The outcomes in both cases can still be treated as either *threats* or *opportunities*. A cost is a threat and may arise from the financial outlay of completing the project, or the financial, reputational, or legal outcomes of not completing the project appropriately or, from adopting the "do nothing" approach. Benefits are opportunities, which fall under the same categories. There are several examples of each which we could apply to our five case studies (see Table 5.2).

Table 5.2 Threats and Opportunities.

Initiating need	Potential costs—threats	Potential benefits—opportunities
Reduction of CO2	Breach of international treaty; loss of international goodwill; increased damage to the climate; increased public anger; large capital outlay.	Security of power supply; public approval; clear demonstration to international community of commitment to climate change.
More office space	Loss of business growth; poor employee retention; poor employee health; capital outlay; loss of reputation among suppliers and customers.	Increased company profile; increased company profitability and value; enhanced employee engagement; improved employee health.
Effective naval deterrent	Loss of defensive capability; loss of life; increased operational costs; large capital outlay.	Improved national security; clear message to potential enemies; improved performance in operational capability.

(Continued)

Table 5.2 (Continued)

Initiating need	Potential costs−threats	Potential benefits−opportunities
Rebrand of equipment	Loss of market share; bankruptcy; poor reputation among customers; reduced employee engagement; financial outlay; company stripped of assets.	Enhanced market share; increased customer perception of brand; facility to improve performance of products; possibility to reduce costs of manufacture.
New range of vehicles	Reduced market share and profitability; poor reputation with customers; bad press; loss of engagement with other manufacturers or suppliers; large capital outlay.	Cooperation with other manufacturers leading to reduced costs; increased market share; facility to use up old components; exclusive company foothold in new market.

It is the function of the client, with the engagement of the key stakeholders, to assess the business case and address all of the relevant costs and benefits of the initiating need, as well as the capability of the organization to engage with it and its capacity to produce it. Capability is in reference to whether the organization has the relevant levels of resources and experience, and capacity is in reference to whether it has the requisite time and physical space to see the project through.

The cost-benefit analysis will help in the shaping of the design at this stage by narrowing down the potential solutions. Of course, it may be that the output of an initiating need may only have one possible solution within the capability and capacity of the organization. In this instance, the business case might be very straightforward, but it will still have practical significance in allowing the client to address the financial, reputational, and regulatory implications.

"Make/Buy" and "Do Nothing" Approaches

A question which is worth asking is whether it is better to make the product in-house or buy it in. The "make/buy" decision also affects those organizations intending to produce the product themselves. It may be that the cost of tooling-up for a new product or product line may outweigh the loss of absolute control in the production stage were the product to be bought in, thereby justifying the investment. Designing for external production will have to take account of potential delays in ironing out issues during the prototype and test phases; and the risk of the external producer lacking capacity to produce the product should also be considered. The input of the procurement or financial team will help to establish cost benefits that are projected many months or years into the future, thereby potentially offsetting large capital expenditure, for example, in the purchasing of machinery or premises. This is the balance between the initial capital expenditure (CAPEX) and the ongoing operational expenditure (OPEX) for the life of the product.

The "do nothing" approach should always be considered in any business case and can be used as a validation of the initiating need—"if we do *not* do this, then *that* will happen." This, too, is inextricably linked to the cost-benefit analysis, where the cost of doing nothing can be weighed against the cost of doing something. If the cost implications to the organization of doing nothing (perhaps in terms of lost revenue) is less than the cost of doing something, over a given period of time, then appropriate decisions can be made. Equally, if the cost of doing nothing far exceeds that of doing something, then the business case can further establish the core constraints of the project: "if we do this, how much will it cost, when must it be achieved by, and what level of quality can we expect."

It is also possible that after completing a business case the client determines that the project should be aborted. Whatever the reason for this decision—lack of financial resources to either fund or maintain the product; aversion to a potential risk; insufficient capacity to produce the output; lack of capability in the organization's team—it should be recorded as the output of the business case for future reference. It is not uncommon for organizations to revisit projects in the future, perhaps after a change of senior management or an injection of capital, or where the risks involved in "doing nothing" have become too great. The information accumulated in a previous business case can prove invaluable in informing these successive decisions.

A useful mnemonic for conducting a business case is BRAUNS, which will help to make sure that the key factors are considered and addressed. It stands for:

- B–What are the likely **benefits** or positive outcomes from pursuing the design?
- R–What are the likely **risks** to be encountered in doing so?
- A–Are there any **alternative** courses of action that should be considered?
- U–Are there any areas of **uncertainty** that could affect the output?
- N–What are the risks, either positive or negative, of doing **nothing**?
- S–What are the **safety** implications of all of these aspects to the organization, the project, and the people likely to be involved?

Key Stakeholder Engagement

Aside from those stakeholders with defined roles and responsibilities in the project, there are likely to be other key stakeholders whose input may help guide the design process. At the business case stage, it may be useful to consider these key stakeholders and open lines of communication with them to prepare for the engagement process. Early engagement with other key stakeholders may also introduce additional elements to the business case that may not have been previously known. For example, in our case study of a nuclear power station, the generated electricity from it would most likely be directed straight into the National Grid who have stringent requirements for power quality and stability. This may require conditioning the electricity generated through a process known as reactive power compensation. This requirement might impact the power station's location, cost, or practicality, and therefore is valuable information for the business case overall.

Responsibilities

At this preliminary stage of the project, where decisions are to be made about the efficacy of the venture, it is important to consider the individuals who will have high-level responsibilities to the project, the client, and the design. But what of those organizations who do not have the necessary skill sets in important areas of the project?

It is entirely possible that a given organization may not have the relevant experience, competency, or availability of personnel to undertake a project to fulfil the initiating need. A small company, for example, which plans to expand its warehousing facility, is unlikely to have a construction expert. In many such instances it is likely the organization will engage external consultants or other advisers to assist them. These external personnel should not be forgotten in the documenting of the project's roles and responsibilities athough it is to be borne in mind that they should always report to a senior stakeholder *within* the organization. This will provide an additional level of assurance that their input is consistent with their contractual obligations.

The business case must consider these aspects in conjunction with the other organizational-level parameters. If the organization is to be too reliant on external advice, this may have cost implications that could make the project unviable. Or, alternatively, the possibility of reduced impartiality or enthusiasm in the project might present as a risk to its timely completion. A lack of experience or availability in the client's own organization must be recorded as a risk and suitable control measures identified. These risks and measures will all have a bearing on the business case as a whole, including the make/buy and do nothing approaches.

At this stage, it may be relevant, and indeed only possible, to identify the following responsibilities.

- The identity of the key decision maker.
- The identity of the key project lead, if different from above.
- The person responsible for creating and/or maintaining the risk register.
- The identity of any individuals who champion any mitigations identified in the risk register.

Once responsibilities have been identified and forms of communication have been agreed, these should be captured in a communications plan.

Design Risk Management

The business case will most likely have considered the make/buy and do nothing approaches, as already discussed. From the consideration of these approaches, and others centred around the financial, prudential, and operational factors of the business case, there may well be a number of risks to the project that are already apparent. Some of these risks may be observable, for example, the potential for cost overrun during the project; while others may be foreseeable, such as whether the market for a new product will still be receptive by the time the product is released for sale.

This moment, therefore, is the ideal opportunity to create a risk register at client level, to document these risks. The register may be relatively light on entries and mitigations in this early stage; this is of no consequence. The raison d'être of the risk register is to record risks throughout the design project by remaining a "live" document and, therefore, being continually reviewed, updated, and added to. By documenting risks at this early stage, controls or mitigation measures may be evaluated further down the line of the design process, but this can only be achieved through a defined and documented system of review, which will form part of the documentation and management system. In the above examples it may be necessary to mitigate the risk of cost overrun by engaging with a particular supplier who can provide materials at a beneficial cost and engaging a marketing consultant to conduct further market studies to ensure that the product will be well-received.

The inherent safety of the project in terms of the core constraints of time, cost, and quality will be dealt with in the form of contracts signed with the producer, suppliers, manufacturers, and other key production stakeholders. This is usually the preserve of the client's procurement department and is the reason why they should be involved during early discussions regarding the project in order to gain their agreement.

The management of risk in general terms begins with the identification of risk within the context and objectives of the output through a communication and consultation process (see Figure 5.2). Once the risks have been identified, appropriate

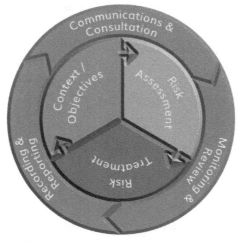

Figure 5.2 Management of Risk.

mitigation or treatment can be applied and suitably monitored and reviewed. The recording and reporting process will help to re-inform the objectives of the output, and the cycle continues.

This cyclical process can be used repeatedly to identify, record, and discuss emerging risk and, through this process, implement effective risk control measures. Of course, there may be risks that cannot be mitigated or controlled through the design or production processes, but the identification of them—and the bringing of them to the attention of the client or end user—will at the very least allow them to be cognizant of them. A client can then decide how they wish to deal with the risk according to their risk strategy. In respect of an end user, residual risks will allow the creation of appropriate instructional or training material. Either way, the objective should be to ensure any residual risk is identified in order to inform the product's subsequent operation, maintenance, and disposal.

At the very start of the design, the level of available risk information may be mostly theoretical. As the design matures, the level of knowledge grows and therefore more informed decisions can be made with respect to the risks and mitigations. Some of the contexts, or areas, of emerging risk that may have an influence of any given design are:

- economic;
- environmental;
- geopolitical;
- societal;
- technological.

All of these influences are to be considered in the risk register where applicable. There are many different formats for registering risk as well as a number of software programs that will help to automate the process. Whilst convenient it may not be necessary to go to great lengths in order to produce a valuable document to control risk and an example risk register is provided in this book (see Appendix A).

An effective risk register should have a number of headings under which detailed information can be recorded as a minimum. These are:

- a unique identification number or code for each risk;
- a brief description of the risk;
- the type of risk;
- the date the risk was raised;
- the risk owner;
- pre-mitigation risk level (frequency × severity);
- planned or proposed mitigation;
- post-mitigation risk level (frequency × severity).

During the design there may be a tendency for some stakeholders to operate their own risk registers: this tendency should be resisted. At the very least, the designer could operate from a copy of the main register from which any risks that are commercially sensitive can be removed. The design risk register is to include the risks to the design process and any proposed mitigations that should be put in place. Agreement is to be sought at the very beginning of the design over the levels of risk that can be managed by the designers, project managers, and organization. This is normally associated with the level and type of risk. The need to rotate a valve, for example, for maintenance access may be a low-level risk and acceptable for the designers to just update the drawings, or it may have a detrimental down-stream impact on the siting of another piece of equipment. The choice of stakeholders involved at each risk review is crucial in order to be able to understand the propagation of risks and make timely decisions over how each is to be managed, controlled, or perhaps possibly flagged upstream of the project.

The business case should consider that:

- The capability and capacity of the organization are available for the project in terms of resources, experience, and knowledge. ☑

- The options of "do nothing" and "make/buy" have been considered. ☑

- The identity of senior responsibilities to the project, including the senior decision maker have been recorded. ☑

- The core constraints that will be imposed on the project from the senior stakeholders and key decision maker(s) have been established. ☑

- The balance between the initial capital expenditure (CAPEX) and any ongoing operational/maintenance expenditure (OPEX) has been validated. ☑

- Any foreseeable or observable risks that are already apparent are recorded in a risk register. ☑

- All pertinent information and decisions surrounding the design are recorded in a suitable system for future reference.. ☑

Requirements Capture

Design Risk Management

Management Structure & Delegations

Documentation & Management Systems

The Four C's

Business Case

Requirements capture

Initiating the design process

The object of a good statement of requirements is to inform the design process but not to control it: it is there to direct, not to dictate. The statement of requirements should encapsulate the initiating need and any outcomes from the business case to allow the designer or design team to present concept solutions based around it. The more restrictive the statement of requirements, the less creative the concept solutions can be, potentially allowing novel design solutions to be missed.

Equally, the statement of requirements should not be so vague as to allow the project to exceed the initiating need and spill over into some divergent solution for which the client made no stipulation. It is therefore important that the statement of requirements strikes the right balance of definition and obliqueness as it can be the difference between a design project that provides the best, safest solution for the client and one which falters or overruns, ending in an output which is unsuitable, unusable, and in need of rework.

It is of course impossible for us here to precisely define a statement of requirements for any given project. Instead, let us look at an example, taken from one of our case studies, and examine the type of language that might be used to inform the statement of requirements based on an initiating need and subsequent business case (see Table 5.3).

Even in this hypothesis, it is hoped the reader can readily determine the completeness of the middle statement: the one that occupies the "Goldilocks zone" of intent, one could say. Equally, it may be readily seen how the statement that is too vague could be easily misunderstood and how design solutions could be arrived at that are supplemental or even superfluous to the original intent. The statement that is too defined is, in contrast, too restrictive in that possible solutions could be repressed or even missed completely in order to satisfy this rather stultifying list of requirements.

The statement of requirements should be carefully considered from the outset and is dependent upon a robust business case being developed, which, in turn, has been derived from a distinct understanding of the initiating need. A well-defined statement of requirements will also provide an excellent reference throughout the design process to ensure that it remains "on track." This is true right through to the validation point of the project where the end product can be held up in a critical light against the statement of requirements to ensure that it has fulfilled the original intention of the client.

Table 5.3 SoR Depth of Information.

Home printer	
Initiating need	The company needs to develop a new home printer range.
Business case	The marketing department report that the existing range is perceived as tired. The existing internal mechanism is readily available and can be supported by the supplier for at least five more years. The marketplace is global. Funding is available.

Possible statements of requirement

Too vague	About right	Too defined
The company requires a new range of printers to update the design of the existing range. The range should have a number of printing options included in each one. Connectivity will be a key element of the company's marketing drive. The internal mechanism should last for five years and the casing materials should be of good quality.	The company requires a range of home printers to match the company's current aesthetic and the latest design trends. Printers should range from one with only printing facilities to progressively include scanning, duplicating, facsimile, and photographic reproduction. All printers should be for home use and able to fit on a desk-top. Accessibility is to be a key element of the company's marketing drive and all forms of connectivity—Bluetooth, USB, Wi-Fi, and so forth—are to be considered. The existing internal mechanism is to be used wherever possible and the casing must include as much recycled material as possible.	The company requires a range of four new printers to match the brand colour scheme and aesthetic. In order, the range will have printing; printing and scanning; printing, scanning and faxing; and printing, scanning, faxing and photo printing. The printers should accept paper between 80 and 100 gsm. Each will connect by Bluetooth version 4.0 and Wi-Fi. Each should fit a footprint of 400 × 400 mm and the casing is to be of 100% recycled materials. The internal mechanism will be supplied externally to the design contract for a period of five years. Power supply will be 230 V AC and the connection will be via a two-pin plug.

The statement of requirements should consider that:

- The initiating need of the client has been clearly identified. ☑

- The right balance has been struck between being informative enough to provide the designer with scope at the concept stage whilst preventing superfluous options or opinions. ☑

- It is used throughout the design process to ensure that it remains true to the initiating need. ☑

- It is used at the validation stage to measure the delivered product against the initiating need and specification. ☑

- The risk register is updated with any further identified risks. ☑

Initiating the Design Process

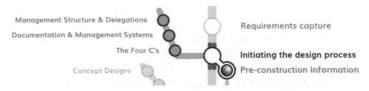

Management Structure & Delegations
Documentation & Management Systems
The Four C's
Concept Designs

Requirements capture
Initiating the design process
Pre-construction Information

Once the client has completed their business case and agreed the statement of requirements with the key stakeholders, the engagement of a designer or design team can begin. It may of course be the case that the designer is part of the client's organization, or is already closely associated with the project. It is entirely possible that the client has sought the opinion or advice of a designer—or even *the* designer—on the matter of the statement of requirements. In complex projects, it may be the case that designers such as materials engineers or civil engineers have been engaged from the outset in order to provide expert advice. Engaging with experts in this way may certainly assist in the definition of the product, and help with technical details that would otherwise introduce delays in the specification or possibly production stages.

Designers and consultants at this stage may agree to operate independently of their colleagues: that is to say, they maintain a virtual barrier between them to prevent potentially unethical judgements. This is particularly the case where large consultancies or firms of architects provide both scoping and design services. The intention is, that, those individuals providing scoping services do not cross-pollinate their ideas with the designers so that design solutions are not influenced by the scoping, and vice versa. This may seem counterintuitive as we have discussed the point of having a well-structured statement of requirements which can direct, but not dictate, the design process. However, it is possible that having a single organization, with a predetermined predilection for a particular type of design solution, providing the client with assistance with the statement of requirements could result in a restricted range of solutions being presented, however inadvertently.

Where designers or consultants have been engaged from an early stage, and where their services may be required during the ongoing design process, it is prudent to examine such advice. This can be achieved in a number of ways.

- Thoroughly examine any advice to establish its impartiality.
- Engage an independent consultant to review the information.
- Review the designer's or consultant's inter-department working policy.
- Review the designer's or consultant's management system.

The output of any design process is entirely reliant on a number of inputs which must be considered in toto (see Table 5.4).

Table 5.4 Design Process Reliance.

Consideration	Reason	Reliant upon
Client's requirements	The fundamental reason for the designed output is to match the client's requirements.	Accurate analysis of the client's concept and provision of a robust statement of requirements.
Financial	To provide an output that is within the expected project budget.	Accurate and defined costings of all of the project's parameters.
Moral	The need to prevent the output causing harm to any person or business during its design, testing, construction, and use.	Fully understanding the design process, testing regime, construction methodology, and user's environment.
Ethical	The requirement of the professional designer to perform their function to the very best of their ability.	The individual understanding the task fully and seeking guidance, without fear or favour, at any opportunity.
Environment	To reduce, to the lowest possible level, the impact of the design output on the planet and its resources.	Accurate assessment of the statement of requirements; understanding of novel solutions; promotion of "green" materials, practices and energy sources in preference to "less green" ones.
Regulatory	To ensure that the output complies with all relevant legislation in the anticipated territories it is to be used in.	Understanding the relevant regulations and their implications.
Safety	The need to provide an output that is safe to be used, in the manner intended, by the persons intended to use it.	Well-informed design specifications that fully understand the environment and manner of use.
Operability	To ensure that the output meets the requirements of the end user in fulfilling their task safely and appropriately.	Well-informed design specifications that fully understand the environment and manner of use.
Maintainability	To ensure that the output is maintainable.	Well-informed design, materials, and machinery specifications.
Disposability	To ensure that the design can be disposed of without impacting the environment.	Well-informed design specifications that fully understand the materials and disposal routes.

It is clear that the design of any device or structure, that is intended to be used by a person or group of persons, must pay regard to not only the *environment* in which that device or structure is to be used, but also the *method of use* too. To fully understand this, the designer must identify not only the correct technical response to the client's requirements but must also understand the operating environment and the methodology of use. This is not to say that the designer must understand fully the operator's job or function. A designer does not need to be able to fly an aircraft in order to be able to understand the challenges faced by someone who does. As project complexity increases, the list of relevant stakeholders also increases, and it is to these stakeholders that the designer must turn in order to qualify any aspects of the design for which operability is critical.

The input of the organization's safety professional(s) can be invaluable here, too. They should be able to provide ample information regarding the correct use and function of devices and structures in use in their organization as well as in their industry generally. They should be able to identify areas that may create greater risk, or where shortcuts in operational regimes have caused incidents or accidents in the past. They should also have risk assessments on safe working practices, which would have been used to inform operators' training and instruction manuals.

This approach we have already called "safe to operate and operated safely." The additional burden of assessing risk and mitigating that risk through the creation of training and instructional models, accompanied by all the relevant supervision, can be reduced radically by a designed output being *holistically* created around the user's needs and function. A fundamental consideration of a design output is to match, as closely as possible, the client's requirements. However, those requirements may not be in the best interests of the final user, and the designer must align these, in full consultation with the client, during the feasibility stage. At this stage, too, the prevalence of cost will become all too apparent in that the client may confuse savings made at the design stage with savings made over the life of the product. It is rarely the case that saving money at the beginning of a project produces an equal measure of savings later on. Indeed, the opposite is almost always the case, where savings made at the project inception invariably create costs further on in the product's life cycle: costs which can far outweigh those initial savings. This is often the difference in the balance between capital expenditure (CAPEX) and operating expenditure (OPEX).

Here again, a competent safety professional should be able to calculate the additional cost of risk mitigation measures for a poorly-designed product over one which reduces the need to rely on, for example, the need for specialist personal protective equipment in the operation of the product. They may also be able to advise on the cost of specialist maintenance procedures or personnel—as opposed to using in-house staff—due to awkward or dangerous maintenance access. All of these costs, when viewed in conjunction with the budgeted project costs will provide a much clearer picture for the client and any financial stakeholders to enable them to make prudent financial decisions.

Management Structure and Delegations

In the publicly available specification PAS 99:2012 (BSI) Specification of common management system requirements as a framework for integration, issued by the British Standards Institute, it states the following with regard to organizational roles and responsibilities:

"The organization should identify, document and communicate the roles, responsibilities and authorities of those involved in the management system and their interrelationships within the organization."

Whether the client in any design project uses a formal management system or not, the need to identify, document, and communicate the roles and responsibilities is vital to the project's success. Early responsibilities will have been identified during the establishment of the business case to ensure that not only the right personnel are available to be assigned to the right roles, but also that the relevant experience, competence, and availability is present. At this initiating stage the identities, roles, and responsibilities of these individuals or groups should be documented. Supplemental information regarding project groups, supervisory and support roles, and signing authorities should also be considered here and documented.

The use of a formal management system can assist in the creation of a robust management structure, but it is not vital. Complex organizations or projects can be well-run from a simple organizational diagram clearly showing the key stakeholders, decision makers, and support personnel. Clearly defined communications routes to key personnel should also be documented so that those wishing to communicate with them, perhaps on matters of design amendments, can direct their query in a timely manner. This is especially important where key individuals are not solely engaged on the project. This is also important where alternates provide cover for individuals who may be away from the project for a period of time, on holiday, for example.

Early assignees to the project will form the key stakeholder group and their influence on the project's outcome cannot be underestimated. It is therefore essential that their responsibilities are clearly defined as well as communicated not only to the other stakeholders but to the wider project team. This may include designers, producers, suppliers, and other members of the client's organization should they become involved, such as operations, maintenance, safety, and so forth.

Within the responsibilities for the roles designated, the following should be considered and the communications plan updated accordingly with the identity of:

- who has responsibility for the transcription of meetings and the dissemination of information;
- who is liaison between the designers, producers, and suppliers;
- who is in charge of the document management system, including the risk register.

Documentation and Management Systems

The creation of the document management system for the project may well be guided by any management system that the organization has in place already, or indeed by BIM if it is mandated for the project. Although management systems, such as ISO 9001, do not formally distinguish a particular type of document environment, the requirements they have for proper implementation may well shape an organization's documentation by inference. Ensuring that the project's documentation aligns with that of the organization as a whole is more important than what that documentation looks like or how it is handled.

Within the responsibilities it may be prudent to identify a "document handler" to, and through, whom all documentation should be addressed. Their role should again be defined by the management structure so that they are confident as to who, when and how they disseminate information. The type and style of documentation should be defined so that all stakeholders are aware that they should be using a particular file type—for example, "portable document format" (pdf)—and that any specific software that is required can be installed in a timely manner, for example, for reading drawing (".dwg") files. Accessibility should be addressed in terms of where the documents are to be held, for example, on a cloud-based server, and who has access to what information. Setting access levels and editing controls is often available with cloud storage solutions.

The issue of copyright and document control can be problematic, especially where individuals engaged on the project may work remotely. These issues should be addressed at this stage in order that breaches of copyright or security do not occur later on. Issues with data protection may be relevant where designers, for example, submit plans or concepts which identify individuals, or where individual's contact details are listed on documentation. The requirements of the Data Protection Act 2018—which implements the General Data Protection Regulations (GDPR) in the UK—should always be recognized in any system used for collecting or transmitting data.

The setting of expectations in terms of interaction with external parties should also be considered at this stage. Is there a requirement for weekly or monthly update meetings, for example? Are meetings to be held on site, remotely, or at head office? What are the anticipated time frames for the delivery of various milestones and what internal discussions or team briefings are to be held? All of them may of course change over time, and certainly in the light of appointing external parties to the project who may bring their own expectations or objections. What is important is that the organization is "setting the bar" as to its expectations and is, therefore, able to decide if changes to these expectations are acceptable or not. It will also allow external parties to gauge whether their own capabilities and capacities are aligned with the client's.

Pre-construction Information

At the initiation of the project, it is also necessary to begin the process of compiling the pre-construction information. This should be compiled by the client's organization or by a competent person appointed by them. Under the Construction (Design and Management) Regulations 2015, this duty is performed by the principal designer if there is one for the project. This information is crucial to the designer(s) and any suppliers of materials or parts as it will provide them with information concerning general and specific risks about the site, location, or proposed installation. This information will be used to ascertain if the design, the materials, or the production techniques are appropriate for the designed outcome. The pre-construction information should begin by recording general information such as:

- the identity and contact details of key stakeholders in the project;
- the anticipated start date and duration as well as the deadlines of any critical phases;
- the organizational arrangement of the project.

Following this, it should list any hazards and observable or foreseeable risks pertinent to the project's location. This may be the site where it is being constructed, the factory where it is being installed, or the environment in which it is to be operated. The pre-construction information is only presenting the hazards and risks for consideration—it is not providing solutions to them. This is an important distinction as it should not be confused with a risk register, which will conceivably list possible control measures. The hazards and risks listed in the pre-construction information are to be *designed-out* by the designer, material supplier, and producer as far as reasonably practicable through the use of the general principles of prevention. There are many considerations for inclusion in the pre-construction information (see Table 5.5).

Table 5.5 PCI Considerations.

Considerations	Example
Administrative and design risks	• Are there any adjoining works? • Has the cost-benefit analysis identified any potential risks? • Are there any specific risks involved in the life cycle or maintenance of the product? • Do any of the proposed designs, techniques, or materials pose any specific risks?
Operational risks	• Are there issues with access, or egress, or security? • Will the installation/construction pose business continuity issues? • Are the operational parameters of the product understood? • Are there traffic management issues with pedestrians and/or vehicles? • Will there be any limitations of use or the need for special training?
Risks to health and safety	• Are there any hazardous materials present or being introduced? • Are there any obstructions or stored energy systems involved? • Are there risks of fire, flood, or other exceptional events? • Are there specific welfare requirements or issues? • Is there to be work in hazardous areas? • Does the manufacturing process itself pose any specific risk?
Environmental risks	• Are there environmental hazards present? • Is there a potential to create an environmental hazard? • Are there regulatory restrictions concerning the proposed installation/construction area? • Are there sustainability issues with the product?

The pre-construction information, like other documents created at this early stage of the project, should remain a "live" document up until the final design stage so as to continue to identify any risks that become apparent during the design stage. Here it will be supplanted by the construction phase plan which is, in effect, the producer's account of how they will deal with any remaining risks that have not been mitigated in the design. The pre-construction information also offers an excellent opportunity to assess the final design against the identified risks to ensure that they have been avoided or adequately controlled by it.

In any event, a designer should not be commencing the concept design stage without the relevant pre-construction information.

Initiating the design process should consider that:

- The impartiality of any advice from designers who have been engaged from the earliest stages of the projects has been considered. ☑

- The management structure to define key personnel, communication pathways, and support roles has been defined. ☑

- Key decision maker(s) and any relevant signing authorities have been identified. ☑

- The type, style, and location of documentation for the project has been defined. ☑

- The document handler, if there is to be one, is identified. ☑

- The pre-construction information has been created and populated. ☑

Design Feasibility

Once the client has agreed the foundations of the project through the business case and the statement of requirements, it is likely at this point that designers will be appointed. This is likely because the client may have the capability to generate conceptual solutions themselves: a car design may be created in-house, for example, with various external engineers and specialists perhaps working with the in-house team later on the technical parameters of the design—engine, aerodynamics, interior aesthetics, and so forth.

If the designer is newly appointed to the project at this point it will be invaluable for them to appraise themselves of the fundamental elements that have led to their engagement. Firstly, what is the initiating need that drove the project in the first place? Is it simply a case of wanting to upgrade an outdated model design or a desire to reinvent the business model? Are there previous designs or products to gauge against? What were the successes or failures they encountered?

If the client is willing to discuss the business case, we can gain from this the parameters of the core constraints. We can judge if the time, cost, and quality balance is achievable, or whether there is a potential for one of them to be sacrificed in favour of the other two. The business case will also provide an insight into the risk appetite of the client, which may prove useful later in the specification stage. It should also allow us to see what range of stakeholders are initially involved in the project and who might be involved as the design progresses.

Moving to the statement of requirements, we should be examining it for clarity and purpose. Has the client specified a "Goldilocks zone" of requirements, or have they been too prescriptive, or too vague? None of this matters if it is truly what the client requires, but a conversation at this point could prevent potential rework and embarrassment later. If the client is unwilling, or unable, to revise or even communicate on the subject, then this can be recorded to demonstrate the designer's intentions later on. It is important for the designer to show that they attempted to prevent rework in order to demonstrate their commitment to providing a professional service to their client and the best possible project output.

The statement of requirements will be supported by the pre-construction information that the designer should be in receipt of before progressing. This will complete the triumvirate of information essential to the designer in helping to shape the conceptual design, namely:

- contract information—this will be by way of tender or contractual document, which will precis the client's business case for the project;
- statement of requirements—covering the specification of the desired output of the project;

- pre-construction information—covering the risks and safety aspects of the construction/installation/production *environment* of the project.

The pre-construction information may help a traditional construction project by, for example, noting the restricted access to the site that could hinder the use of large, prefabricated components. In an engineering project, it may note the client's extensive CNC (computer numerically controlled) manufacturing capabilities (which the business case has asserted should be utilized in the project) and, therefore, any designs requiring an alternative production method—such as selective laser sintering for 3D printing—will add set-up costs to the project.

The designer must also understand the management of the project, from the client's viewpoint, and be cognizant of the role they (the designer) have to play in it and who, from the client side, is going to be pivotal in day-to-day communication, decision-making, and project management. The designer should also be reviewing the document management system to see how this integrates with their own, if necessary. Are there potential pitfalls to the document types, formats, locations, or access requirements? Resolving these at this stage will prevent errors and omissions occurring later.

Only when the foundations of the project have been reviewed and agreed on can the process of creating concepts for consideration begin. The objective of the feasibility stage is to provide solutions to the question posed by the statement of requirements. At this point, very little should be considered "off limits" in terms of possible design solutions, even if they seem initially at odds with the information made available to the designer discussed so far. We should accept that if, as a designer, we provide a concept that is outside the remit of the business case, the statement of requirements or the pre-construction information, we are likely to encounter resistance from the client. If there are proven grounds to suggest such possibilities, however—for reasons of cost, quality, time saved, safety, or best environmental practice—then it is right to make such a suggestion, especially if it can be supported by strong evidence.

The reasons for making these suggestions are twofold. Firstly, the designer has a legal duty to improve safety wherever practicable and an ethical duty to provide their client with the best possible solution. Secondly, the designer may be familiar with solutions that the client is not aware of and this may steer the client in a slightly different direction to what they had originally envisaged. The provision of novel or unorthodox solutions is dependent on the following caveats.

- The designer should be aware of the information contained in the:
  - project contract or tender document;
  - statement of requirements;
  - pre-construction information.
- The designer should be aware of the effect of the core constraints balance (i.e., affecting either time, cost, or quality will affect the other two either positively or negatively).
- Any proposed concept should be considered in regard of safety (along the lines of the general principles of prevention).
- Safety can *never* be compromised by savings on cost.

Environmental and External Influences

We have so far discussed the initial influences on the design project, which are:

- the client's requirements;
- the core constraints of the project;
- the ethical demands on the designer and key stakeholders;
- the promoting of safety throughout the process.

All of these influences will have been identified and recorded in the information assimilated as part of the project's initiating procedure. The designer will therefore be cognizant of them and should be addressing them within the remit of their design and the management of the design process in general. In addition, there will be further influences that become apparent during the design process, particularly if and when more stakeholders become involved at the functional level. These can be grouped as:

- operability;
- maintainability;
- disposability.

These influences will help to shape the specification and final design stages of the process as the design matures. Although the designer may have these influences in mind during the concept stage, it would be obstructive to consider them in detail as it may risk missing a novel concept suggestion that succeeds on every other level. These operational influences can often be addressed later in the specification stage. Influences outside of the direct control of those involved in the project include the following:

- public and socio-economic;
- political (including geo-political).

These factors can be subject to dramatic change, even after long periods of relative stability. It is readily accepted that perhaps few design projects would need to take into account these influences, with perhaps only large, complex, or infrastructure projects, or possibly those concerned with breaking into emerging markets being affected. And major shifts in public opinion or political will may be difficult, if not impossible, to predict. With the increase in international trade and organizations with operations in multiple state territories, however, it is important to ensure that signs of shifts in these influences that may affect the project are not discounted.

There are two immutable influences that the designer cannot ignore. These are:

- environmental;
- regulatory.

Environmental influences may be those considered to have a small "e"—that is, those that are concerned with the location the product will be used in; and with a capital "E"—that is, those concerned with the wider environmental concerns that have become ever more prevalent and pressing. Influences in regard to the locations in which a product will be used relate to matters such as:

- If the product is being used in a volatile atmosphere, is it proven against releasing chemical agents, heat, or electrostatic sparks, for example?
- If the product is to be used at height, can it be fitted with a lanyard to prevent it being dropped?
- If the product is to be used in extreme conditions, is it proven against the likely range of temperatures, pressures, humidity, as well as the ingress of dust?
- If the product might be used over or near water is it waterproof and buoyant?

In relation to wider environmental concerns, the effects are akin to those of social and geo-political influences. Governments are under increasing pressure (and rightly so) to deter behaviour that impacts our planet: note the recent bans on plastic granules used in skin care products and plastic stems in cotton buds. Societal influences can also affect, and be affected by, environmental concerns, such as the reciprocation of supply and demand in connection with electrically-powered cars. "Green" design may still be more connected with marketing than output at the current time, but the journey to sustainability must begin somewhere and the reciprocity with public demand, as in the example of electric cars, can only yield positive results. The authors of this book certainly recommend the reader to consider the wider environmental impact of any of their projects with the same level of concern that they would consider safety.

Regulatory influences of course may well be connected to either type of environment discussed above, as well as to other criteria such as the workplace, the intended use, material specification, and so on. The requirement for environmental impact assessments, for instance, has raised the consideration of product disposal very much to the fore: motor manufacturers having to consider the recyclability of their cars during the design stage is such an example. In the UK, the Provision and Use of Work Equipment Regulations 1998 (PUWER) has requirements for the supply, use, maintenance inspections, and disposal of work equipment and, although these regulations are in response to a European Directive (2009/104/EC)—and should therefore apply uniformly across the European Union—different countries apply the Directive in slightly different ways.

Thus, some designs could be constrained in some respects by regulatory influences such as PUWER in the country of manufacture, and in other respects by the local interpretation of the same regulations in the intended destination country. Added to which, there will be Environmental influences to consider.

The level and complexity of the influence that external factors have on the design will of course be entirely dependent on a multitude of parameters, many of which will be detailed in the triumvirate of information the designer receives prior to setting to work. To visualize the comparison of influence that these factors can have, let us consider four specific types of design, A to D (see Figure 5.3).

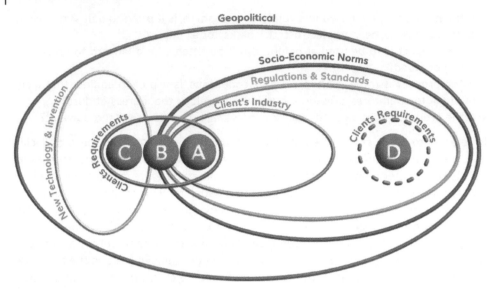

Figure 5.3 Design Scenarios and Risks.

Design A

This type of design is one that sits well within the client's area of expertise or operation and will have well-established applicable regulations and standards. These will form the governing framework for any proposed design solutions. There will also be accepted norms for this type of design in terms of the industry sector as well as the wider public domain. It may be considered low risk as far as the client is concerned, in terms of their business case, as they will understand what the product should look like and how it should be operated, maintained, and finally disposed of.

Unfortunately, this may be the type of design where the designer can be least creative because of the large amount of preconception surrounding it. Implementing sustainability into the design may also prove problematic for similar reasons and yet, paradoxically, this is perhaps the one that requires most involvement in terms of considering the circular economy due to the widespread use of this type of product.

Design B

This type of design sits mostly within the client's area of expertise but may contain certain aspects that involve new technology and/or invention. This might be as-yet unproven cutting-edge technology, or perhaps technology that is not yet widely accepted. It may even be of a type that the client or designer is themselves developing or promoting. The risks involved here will be greater due to the inclusion of this new technology which, we must presume, will be integrated in some way with established technology or principles. Thus, there would need to be a greater period of testing undertaken.

Risk influences from such things as the supply chain, regulatory, socio-economic, and geopolitical impact will take more consideration in the risk register due to uncertainty among trade, national markets, and the general population. Ironically, with this type of design, it is probably easier to implement new sustainable ideals due to the lack of precedents in the market.

Design C

This type of design sits firmly in the "unknown" area of technology and invention. An example would be the first manned space rocket. When the initial research into space flight was underway, no one had any information on the survivability of humans when subjected to enormous G-forces and weightlessness. There was no data on the risks involved with launching, orbiting, and re-entry through the Earth's atmosphere. As it had never been attempted previously, there was no societal norm driving the requirement for legislation or standards. Despite the extremely high risks, it was felt they were outweighed by the possible advances in knowledge and further invention. Therefore, there was an acceptance of the costs and level of effort that would be required to minimize these risks to acceptable levels. It is important to note that the "acceptable level" of risk associated with early space flight was perhaps at a level that we, today, would be uncomfortable with.

This type of design creates potentially the greatest opportunity for implementing thought-provoking and possibly promising advances in techniques, materials, and sustainability, although clearly at quite an elevated level of risk in terms of reputation, financial impact, and of course, safety.

Design D

This design reflects a product type that is perhaps well-established within the client's own sphere of industry or operation. The most fitting solution to the client's need may be based on one that other organizations in the same industry, or perhaps other closely-related industries have already researched, developed, and resourced. The client's precise needs, however, might require this to be adapted but, on the whole, this may well be considered a low-risk solution. An example might be the design associated with that of wind turbines where the well-established technology of aircraft propellers from the aeronautical engineering industry has been utilized on a grand scale.

Despite the possibly wide-spread acceptance of previous similar products, this type of design could provide excellent opportunities to improve techniques, material use, and concepts of sustainability as it is riding on the back of much current thinking and levels of acceptance and yet could, in its own right, blaze a new trail in these fields.

General Principles of Prevention

The general principles of prevention exist to reduce the risks associated with any number of hazards and can be readily applied to any risk environment. As we have discussed, a hazard is something with the potential to cause harm, and risk is the quantum of likelihood and severity that harm may occur. A hazard can be a physical thing, such as electricity, or it can be an intangible thing, such as a financial investment. We have also noted that risk can have both positive and negative outcomes. These we have described as opportunities and threats.

The regulatory concern with the use of the general principles of prevention is in relation to identifying, avoiding, and controlling the risk of harm to individuals. But the hierarchy of the principles can be equally applied to financial, marketing, supply, and management aspects of a project, as well as to the project as a whole. Let us consider how these principles might have been applied to the business case of a project, for example, (see Table 5.6).

Table 5.6 GPoP Applications.

Principle of prevention	Possible application
Avoid risks	Ensure the project is within the capability and capacity of the organization. Ensure financial reserves are robust enough to meet possible overspend. Confirm the competences of all third parties involved with the project.
Evaluate risks that cannot be avoided	Create a risk register to monitor observable and foreseeable risks. Examine external influences to establish possible risks.
Combat risks at source	Hire in expertise where the organization's capability is weak. Take out project insurance.
Adapt the work to the individual	Ensure the best placed personnel are given the jobs most suited to them. Hire in expertise to train personnel if more projects are to be undertaken.
Adapt to technical progress	Engage with suppliers, designers, and engineers to establish if there are more suitable alternatives to materials, designs, or techniques for the product.
Replace the dangerous	Establish a project management system to suitably control the project. Ensure compliance with the CDM regulations through competent advice and audit. Establish appropriate contracts with suppliers. Engage a wide range of potential stakeholders (operators, maintainers, etc.) to comment on the project's output.
Develop a prevention policy	Develop a communication and management policy to control the project. Ensure additional funding is available for possible design changes. Control scope creep through a robust statement of requirements.
Provide protective measures	Ensure engagement from senior stakeholders and decision makers throughout the project. Engage professional independent advisers. Take out insurance against external influences.
Provide instruction	Disseminate the project management plan to all relevant stakeholders. Ensure role holders within the management structure are aware of their responsibilities.

From this, we can see that the principles can be used as a tool to promote a safe environment for any aspect of the project. It may be that some of the principles might not apply in all cases: this is no matter. What is important is that each element of the hierarchy *is considered in any event*, and returned to as often as necessary to ensure that no emerging risk has been left uncontrolled or without consideration. The measures considered in each application of the general principles of prevention should be recorded and kept with the project's documentation.

As an example, let us look at how these principles can be applied to a hypothetical design of a complex industrial machine with which a number of people, having various skill sets, will interact (see Table 5.7). These people will include operators, cleaners, maintenance personnel, and setting engineers. For clarity we shall ignore the building in which the machine is housed. For each principle, we have recorded a possible control measure aligned with the eliminate-reduce-control hierarchy. The resultant output would, of course, depend on many other considerations of the design project, that is, cost, client's requirements, environment, and so forth.

Table 5.7 GPoP Control Measures.

Principle of prevention		Possible control measure
Avoid risks	Eliminate	Automate or robotise hazardous processes in order to exclude operators from the vicinity.
		Mount sections that require maintenance—e.g. motors and gearboxes—on slide-out trays to prevent reaching in.
	Reduce	Manufacture the machine in several parts to reduce risks with installation and final disposal.
	Control	Locate sections requiring maintenance behind fixed guards interlocked to the power supply.
Evaluate risks that cannot be avoided	Eliminate	Conduct robust design reviews with all relevant stakeholders to establish risks and agree elimination measures.
	Reduce	Minimise or rotate exposure to risks through work rota.
	Control	Implement and record a training regime.
Combat risks at source	Eliminate	Use, hard-wired, wireless and battery powered systems.
	Reduce	Route trailing leads in floor channels so that they are not on the floor.
	Control	Embed trailing leads in rubber strips

(Continued)

Table 5.7 (Continued)

Principle of prevention		Possible control measure
Adapt the work to the individual	Eliminate	Automate repetitive tasks.
	Reduce	Provide sound proofing around areas of the machine where operators work.
	Control	Use shift rotation to lessen the exposure of operators to emissions or psychosocial hazards (repetition, fatigue etc.)
Adapt to technical progress	Eliminate	Use high-speed cameras and computers to analyse product instead of operators.
	Reduce	Use predictive algorithms and monitoring software to predict maintenance requirements.
	Control	Implement a programme of planned preventative maintenance (PPI).
Replace the dangerous	Eliminate	Replace conveyor belts with air beds or roller beds.
	Reduce	Encapsulate conveyor belts to prevent pinch points and drawing-in risks.
	Control	Install lock-out tag-out (LOTO) mechanisms on hazardous components to prevent unauthorised access and inadvertent operation during maintenance or cleaning.
Develop a prevention policy	Eliminate	The policy requires the use of automation in areas of elevated noise levels.
	Reduce	The policy specifies the buying of machinery that emits noise below the level at which hearing protection is required.
	Control	The policy requires the wearing and supervision of hearing protection.
Provide protective measures	Eliminate	Isolate or enclose hazardous processes of the machine to negate the need for operators to use PPE.
	Reduce	Install local exhaust ventilation systems to remove dust emissions from an area of the machine.
	Control	Implement policy, training and supervision systems to ensure operators wear appropriate PPE.

Design Review—Feasibility

The purpose of the design review at this stage is to confirm the feasibility of the intended design output by assessing the proposed concepts against the background of information pertinent to it: business case, statement of requirements, and pre-construction information. Additionally, there should be posed such fundamental questions as:

- Is the intended design feasible?
- Is the intended design as required?
- Can the intended design be technically produced and in the required quantity?
- Is there a marketplace for the intended design?
- Are there any limitations of use for the intended design, and are these acceptable?
- Has the organization got the capability and capacity to produce the intended design?

From the business case, the design review should be assessing that the relevant financial input remains available for the project, especially if any of the proposed concepts have taken an approach relying more on initial capital expenditure (CAPEX) to significantly reduce lifetime operational expenditure (OPEX). This is where having the early engagement of the client's financial/procurement teams can prove invaluable as they may well be able to provide timely answers to these questions. The proposed concepts should be reviewed against the core constraints that the client has set in place and, again, having the appropriate decision makers in conference at the time can produce swift answers.

As well as the initial financial input required in the project, it is important to understand if any proposed concepts, particularly those of a less orthodox nature, present an operational expenditure greater than that anticipated. Savings at the beginning of a project do not always reflect in savings overall during the product's lifetime, thus requiring a balanced decision on the costs and benefits to be made. This can be equally applied to the validation of the product prior to handing over to the client or end user. The validation, where the product is tested against various parameters of its functionality, can prove to be costly or time-consuming if complex systems have been introduced during the design process. These costs in time and money may of course be repaid if the systems offer, perhaps, operational savings during the life of the product.

The statement of requirements should be reviewed against the proposed concepts to ensure they meet this specification as fully as possible. Variations may be due to the matters raised above and, once more, engaging the appropriate stakeholders at this juncture can provide the answers as to whether the client is prepared to accept these variations.

The feasibility design review is an opportunity to test the proposed concepts against the statement of requirements. This is to ensure that:

1. The client's statement of requirements remains valid with regard to their original intent.
2. The designer has the right quantity and quality of information to progress the design further and has considered this information appropriately.
3. Emerging risks connected to the design, production, operation, maintenance, and disposal of the product are identified.
4. Variations to the time, cost, or quality of the product can be analysed and either agreed or disallowed.
5. Novel or unorthodox proposals can be discussed openly and analysed in terms of lifetime costs, benefits, and suitability.
6. Senior stakeholders and key decision makers have an opportunity to review and refine the design as necessary.
7. Approval to move to the next stage of the design process is formally documented.

Finally, the pre-construction information should be reviewed against the proposed concepts to ensure that any relevant implications of constructing, installing, or assembling the product are suitably addressed. Where any risk remains, perhaps where operability may be constrained in some way, this should be documented in the risk register in order that it can be addressed as required, either during the final design, construction, or in-service phases.

Once these reviews have been conducted as described above the outcomes should be recorded for future reference, particularly in the subsequent design reviews where this information can be cross-referenced and updated. The outcomes should be arranged under the following headings:

- design risks;
- construction/production risks;
- operational risks;
- maintenance risks;
- disposal risks.

Whilst not all of these sections may be able to be completed at this stage, it will at least allow the designer to see, at a glance, any areas of the design that require particular attention. It can also be used to record any variations that have been agreed, or where proposed variations have been disallowed and the reasons behind this. This information is invaluable in cases where, for example, key members of the client's team are replaced and incoming members of it query decisions made prior to their appointment.

At this and each subsequent design review, there are four possible outcomes.

- Pass—the design is maturing as expected and is at a level sufficient to pass on to the next stage of the design process.
- Pass with actions—the design is maturing as expected with the exception of specific deliverables or identified risks that must be resolved within a given timescale.

It is still possible to progress on to the next stage of the design process with caveats.

- Fail with actions—key deliverables in the design have not been met or there is too great a risk posed to progress at this stage at this time. A timescale should be agreed on actions to take and a date set for a repeat design review (which may be an abridged version).
- Fail—too many deliverables in the design are missing, or risks to the project have been identified that are insurmountable. The design cannot continue in its current iteration.

The formal approval to pass the design allows the designer to move on to the specification stage with the knowledge that their client has witnessed and agreed all the salient information the designer has provided for the proposed solutions. Any alterations after this point by the client can be addressed appropriately in terms of expense to the project, and to the designer. Not having this certainty can create friction where late alterations or amendments arise and the reasons for them cannot be readily pinpointed.

The feasibility design stage should consider:

- That any environmental and external influences that may affect the design, production, operation, maintenance, or disposal have been clearly identified. ☑
- Initial estimated costs so that the various concepts can be compared are included. ☑
- How the product is to be created, operated, maintained, and disposed of. ☑
- Concepts that are novel or unorthodox if they provide practical or pragmatic solutions within the remit of the statement of requirements are not excluded. ☑
- The advantages and disadvantages of each concept solution are highlighted. ☑
- How risks have been, or can be, reduced through the use of the general principles of prevention is demonstrated. ☑
- A feasibility design review, prior to moving to the next stage, is conducted to agree the design(s) to proceed to the next stage. ☑
- The risk register is updated with any foreseeable or emerging risks that have been identified. ☑
- Acceptance of the design is formally documented. ☑

Additional Stakeholder Engagement

Once the feasibility of the proposed concept has been tested and a design solution agreed upon, it is prudent to then engage with additional stakeholders who will have an interest in the product. These may be individuals or groups who will:

- supply materials or parts to the production;
- maintain or repair the product;
- operate or control the product, or supervise those that do;
- oversee its operation.

This final group should include safety engineers, occupational health and safety practitioners, operation managers, and human resource managers.

The reason for engaging with this group is twofold. Firstly, those who oversee the use or operation of a product in the field can have valuable insights into its practical application including any perceived or actual limitations or restrictions during its use. Secondly, they may have evidence—either recorded or anecdotal—of previous issues, incidents, or accidents related to a product's use, operation, maintenance, or repair. Human resource managers may be able to provide information about work or procedures through exit interviews with workers or sickness records.

All of this information is valuable to the designer in providing operational information. It can also help to eliminate rework later on because, perhaps, a switch was put in the wrong place, or a door couldn't open due to an obstruction, or a valve could not be accessed by the maintenance engineer. These may seem trivial matters but these and similar issues have been the cause of much expensive rework in the past. Maintenance and repair engineers are especially important in the engagement process as they are the ones who regularly interact with products that are operating outside of their normal parameters, and this is significant. We suggest there are three states of operation, as follows:

- **Normal.** The product is within its operational parameters and functioning as specified. The intended end users interact with the product as they are trained/authorized/permitted to do and normal safety protocols are as intended.
- **Abnormal.** The product is not within its normal operational parameters. Safety protocols may be disrupted or bypassed (e.g., guarding removed from a machine, bow doors open on a ferry, fire doors removed from an office). The intended end users should not be permitted to interact with the product unless they have received additional and suitable training/authorization/permission to do so or additional appropriate supervision is provided.
- **Exceptional.** The product has failed to operate normally: for example, the machine has seized up, the ferry has started sinking, the office has caught fire. All intended

end users, including maintenance and repair engineers, should be removed from the vicinity of the product. Emergency protocols now supersede any others and only those with specific training/authorization/permission should interact with the product.

All products are subject to abnormal operational conditions at one time or another, whether that is due to maintenance, repair, cleaning, refitting, painting, replacement, or adjustment. The designer should consider these during the feasibility stage to establish whether they are potentially controlled or whether there needs to be further development of the design to overcome any issues or, indeed, whether the design cannot be progressed as it is. Hence the need to engage with those stakeholders who are destined to interact with the product when it may be in an abnormal state in order to ensure that safety remains paramount.

Exceptional or emergency conditions may not be factored into the design requirement: very few people regard the possibility of their newly-developed project suffering a catastrophic failure. But failures do occur and it is incumbent on the designer to at least consider the possibility of the very worst situations. Such considerations might include the following.

- Stored energy systems: Are they designed to de-energize in the event of a major failure?
- Pipework: Is it fitted with relief valves to prevent over-pressurization?
- Fire: Can it be contained by material choice, location, or other physical factors?
- Electricity: Is there a remote kill switch? Is it connected to heat or smoke sensors?
- Explosible atmospheres: Are systems designed specifically to operate within them? Are leaks designed to be detected and can they be isolated promptly or automatically?

Often, simple design measures encapsulated at this early stage can not only prevent costly rework later on, through additional safety devices and control measures having to be installed, but may also reduce the potential harm to people and property in the event of a major or catastrophic failure. Paying regard to the pre-construction information will also aid the designer in appreciating the intended operational environment of the product as well as any limitations or restrictions imposed upon it.

Supplier Engagement

Materials engineering has developed almost exponentially over the last few decades, and continues apace. Suppliers of materials or components to the subsequent production stage of a product can provide valuable information about novel or cutting-edge articles that may well aid the designer in providing ever better solutions. But cutting-edge materials and components may have experienced only limited functional use in the real world and, therefore, should be subject to robust testing and investigation before being committed to. Competent suppliers should be well-placed to help with this process.

Even everyday items that are destined to be specified in the production process can be subject to the vagaries of supply and demand and logistical problems. Again, engaging with prospective suppliers will aid the project by informing the designer of possible supply chain shortfalls. In complex designs, it is often the case that the final product will be delivered in phases, with work on each subsequent phase dependent on the finalization of the one before. Equally, many projects employ time-management techniques such as LEAN, just-in-time, or Six Sigma. These techniques can demand the arrival of critical components at pre-determined critical times and failure to meet these demands can jeopardize the entire project. Confirming the availability of parts and materials well in advance will provide a better project workflow, produce a higher quality product, and meet the financial expectations of the client.

Obsolescence management at this early stage will question whether your chosen component will be available throughout the designed product lifetime, or whether there need to be allowances for the use of other products in the production of the design, or as replaceable parts during the in-service life. A laptop designer isn't worried about whether the processor can be upgraded and therefore hardwires these in. They do, however, consider memory upgrades and therefore allow for access and replacement. The suppliers of memory (RAM) support standard pin layouts so that it is backwards compatible.

User Requirements

We have discussed the benefits of engaging with those groups or individuals who oversee the operation and use of a product. Of equal importance are those who actually perform the interactions with the product: the end users themselves.

It would be churlish to expect a designer to hold court to a seemingly endless procession of machine operatives or maintenance engineers to gauge their considered opinions. In general, workplaces operate on a stratified system: for each team of workers there will be a team leader; for each group of team leaders there will be a supervisor; for each group of supervisors there will be a manager; for a given number of managers there will be a senior manager or department head, and so on. Therefore, depending on the size, complexity, and output of the project, it should be relatively straightforward to engage with an individual who represents a number of opinions.

Workshops are one method of engagement that have proven useful time and again. Short presentations of the intended output of the project followed by brief question and answer sessions can help to allay concerns as well as garner valuable insight into preferences, wishes, and any operational issues that have perhaps eluded the management team. Let us look at some of the possible insights that could be gained from additional stakeholders who might be involved with projects connected to our case studies (see Table 5.8).

Table 5.8 Additional Stakeholder Insights.

Project	Additional stakeholder insights
Home printer	• Sales department: returns due to failures, breakages, quality issues, packaging. • Marketing department: feedback from distributors, customers and product test groups; reputational issues from social media. • Suppliers: news of changes to materials and components; issues with regard to supply lines and logistics. • Field engineers: warranty issues; repair techniques including workarounds.
Nuclear power station	• Maintenance engineers: access issues; breakdowns and failures; false positives in warning systems. • Safety practitioners and engineers: minor incidents and accidents. • Human resource managers: sickness records, including wellbeing.
Office block	• Maintenance engineers: cleaning of windows, access to pipework, replacement of fixtures and fittings. • Electrical and electronic engineers: access to cabling and trunking; provision of fire points, power points, data points. • Fire engineers: issues with materials, especially in combination; egress points and pathways; the psychology of people in stressful situations. • Office and human resource managers: worker perceptions, preferences, and overall wellbeing status.
Warship	• Maintenance staff and asset management records to provide information on running times, failure rates, and so forth. • Command staff to provide feedback on operational scenarios. • Operators to provide feedback on systems/equipment performance in full operation (peacetime and in theatres of operation). • Aircrew (pilots/navigators) to provide feedback on aircraft operations from vessels. • Special forces to provide feedback on operational support.
Motor car	• Dealer network: issues with warranties, recalls, and so forth; brand reputation; customer expectations and preferences. • Marketing department: customer feedback; media credibility. • Suppliers: advances in materials and components; supply line issues; paint colour trends.

Additional stakeholder engagement should consider that:

- Any prospective suppliers are engaged to discuss how materials or components might inform or affect the specification and production stages. ☑

- Any user requirements for normal, abnormal, and emergency operations of the product have been identified. ☑

- The risk register is updated with any foreseeable risks during all three states of operation. ☑

- All relevant parties are included. ☑

- The management and communication structures are informed about the amount, type, and frequency of additional stakeholder engagement. ☑

Design Specification

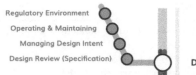

Regulatory Environment
Operating & Maintaining
Managing Design Intent
Design Review (Specification)

Design Specification

The specification design stage is where the chosen design from the feasibility stage should be "made real." In this sense, it is the critical point where the design is tested against the actual environment, both physical and legislative, in which it will operate and against the operational expectations and parameters.

The concepts have now been presented and the outline for the design has been reviewed against the client's requirements, culminating in an agreed design. The specification stage adds detail to that design to enable it to be tested or possibly modelled to prove that it will work and still meet with the customer's expectations. During this stage there may well be significant questions and findings that may push the envelope of acceptability.

The specification design stage should involve as many additional stakeholders as are necessary to continue to inform the design process. Although the number and range of these stakeholders should have been decided early in the project and documented, it is important to remain vigilant for any stakeholders that may have been missed, or where it becomes apparent that the input of previously neglected stakeholders would be valuable.

The designer should be wary of errors that can be caused by the client amending or adding to the specification during this stage, causing the scope of the design to "creep" from its original ambition. "Scope creep" is to be avoided if the project is to remain relevant to the statement of requirements as this can be deleterious to the design process, not to mention a potential waste of time, money, and effort in the project. Maintaining accurate records of discussions and transactions between client and designer is crucial in alleviating this problem.

This stage is also an opportunity to ensure that any recommendations or actions recorded in the previous design review have been considered. This will include any testing or prototyping that may need to take place in order to substantiate a particular element or system of the design.

Regulatory Environment

We have discussed how the regulatory environment acts as an external influence on design and, as we progress through the specification stage, we should be confident that overarching regulation has been at least considered thus far. There should also be the confidence that the designer has begun the process of applying the general principles of prevention and, therefore, that risks have at least been identified if not already mitigated or controlled.

During this stage, further regulation may well come to light, and the designer is required to be aware of this. Such regulation may be involved with:

- the supply or use of particular materials;
- specific importation requirements for materials or components;
- the specification of certain components, either in their manufacture, use, or design;
- the interaction of people and the product, that is, accessibility, ergonomics, noise, emissions, and so forth;
- the control of energy sources either into, or from, the product;
- the control of emissions from the product.

It is quite likely that many, if not all, of these regulatory considerations have already been factored into the design. A project to design a printer, for example, will already have electricity and CE marking regulations in mind. But what of a building that is designed to have a certain type of exterior finish that, due to some previous event, becomes unavailable in the home market, or is banned in the intended destination market. Only some two weeks before the United Kingdom was due to leave the European Union's free marketplace there was still no trade agreement in place, meaning that companies on both sides were not aware of what regulatory environment they would be buying from or supplying to.

Foreseeable risks around the regulatory environment should have been recorded in the risk register at the beginning of the project. Additionally, foreseeable risks surrounding the project in terms of regulations that *may* come to apply to it should have been recorded in the risk register. It is far better to have spent a small amount of effort on considering the possible legal implications of a product early on, rather than discover as the silk drape is about to be removed from your beautiful new product that it has just become illegal in the destination market.

Operating and Maintaining

We discussed in the additional stakeholder engagement section the relevance of gauging the opinions of those who would ultimately interact with the product, either directly or indirectly. Whilst no opinion should be arbitrarily regarded of greater merit than any other, those who will operate and maintain the end product will, by virtue of their interaction, have arguably the most relevance. Operators, because they will interact with the product for the greatest period of time during its normal operation, and maintainers, because they will potentially interact with the product during abnormal operation; perhaps when the product is in maintenance mode, or key components have been removed or exposed. Safety legislation requires us to consider the greatest risks first, whether that is due to the severity of the risk (the amount of harm it could cause) or the likelihood of the risk occurring (the frequency of the risk or the length of time the risk exists).

The requirements of operators and maintainers may, at times, be conflicting, due to the differences in normal and abnormal operational parameters. In such cases, neither set of requirements should take precedence arbitrarily but should, instead, be assessed for risk as usual and with regard to the calculation of *likelihood × severity*. The additional skills, experience, and training generally required by maintenance personnel should also be factored into this calculation. What must not occur, however, is the conscious decision to exclude maintenance safety for reasons of cost.

Aesthetics often play a key role in product design. In our case studies, the design of the home printer, the office block, and the car are likely to be motivated quite highly by the look of the product and how it is perceived by the end users. But here, too, the importance of safety in abnormal operation conditions must be observed where aesthetic qualities lead to the unusual or unorthodox placing of maintainable components. This could be achieved through the use of interlocks, specialist training or the resources needed to complete any particular maintenance task. The scheduled time for Bugatti technicians to undertake an oil change on a Bugatti Veyron, for example, is 27 hours due to the fact that virtually the entire rear end of the vehicle must be removed to access the sixteen oil drain points and all the filters. This is a methodical approach to ensuring the safety of the technicians as well as the driver of a very high-performance car. Clearly, the owner of any car costing several million pounds is unlikely to baulk at a service costing tens of thousands. This would clearly be untenable in a much less expensive car.

In a completely different realm, the development of software has, over recent years, often become a race to update and beautify at the expense of user operability. Implementing change for the sake of change, or because of an aesthetic whim, may be prized in design departments but is unlikely to win support, particularly where software or systems are widely popular or extensively used in business. The dramatic changes to the suite of office software from Microsoft in 2007 (the introduction of the "Ribbon"), which caused widespread consternation among people who had grown up with a tried and tested format, is a case in point.

Let us examine some of the possible influences that operators and maintainers might have on the *specification* of a design. These influences may result from active engagement, focus groups, design workshops, or perhaps from historical experience with similar products or environments (see Table 5.9).

Table 5.9 Operator and Maintainer Influences.

Product	Possible operator influences on specification	Possible maintainer influences on specification
Nuclear power plant	Alarms and higher priority levels overriding lesser priorities. Positioning, labelling, and types of human interfaces. Protective covers on specific switches. Failsafe devices.	Access requirements. Maintenance envelopes. Working way to remove equipment. Isolation locations and access. Radiation zoning.
Office block	Style or type of finishes. Layout of floorspace. Type, location, and quantity of cabling (data, power, etc.). Air quality. Lighting (quantity, colour temperature, etc.).	Access to power and data cabling. Access to HVAC. Access for regulatory inspections and maintenance.

(Continued)

Table 5.9 (Continued)

Product	Possible operator influences on specification	Possible maintainer influences on specification
Warship	Bridge and control room layouts. Command desk functionalities. Communications. Fall-back modes of operation.	Access requirements. Maintenance envelopes. Working way to remove equipment. Isolation locations and access.
Home printer	Functionality. Location of primary controls. Type of material surfaces (with regard to cleanliness, feel, aesthetics, etc.).	Access to change inks. Access for paper insertion and automatic paper alignment. Cleaning access.
Motor car	Availability of optional extras. Seating layout. Availability of colours.	Accessibility to service items. Jacking points. Types of fixings used. Parts compatibility with current vehicles.

Design Review—Specification

The purpose of the design review at this stage is to confirm that the specification of the intended design output meets the expectations of:

- the client's design intent (from the statement of requirements);
- the regulatory requirements for the product, its materials and components, and the environment in which it is intended for use;
- the wider stakeholder group.

In short, the specification design review is primarily concerned with the question "is the design fit for purpose"? Issues with the design at this stage can be readily rectified prior to committing to the production stage, thereby saving potentially expensive rework costs.

The design review should be attended by the key stakeholders as well as the key decision maker(s). It is also valuable to request the attendance of the procurement department, or financial decision maker, if they are different to those previously mentioned. The presence of these key role holders at the review can help to ensure that where the design has been subject to scope creep (be that in terms of design, specification, or cost) then decisions can be made in a timely manner as to whether it remains within the client's original intent. The evidence for and against any deviations can be discussed in real time, leading to well-informed decisions that all key stakeholders will be privy to at the same time. A robust communications plan will provide the framework here for the identities of these individuals and the expectations they can have for information flowing to and fro, in terms of emails, meetings, and so forth.

At this review, all outcomes from the previous (feasibility) design review should be revisited. Any outstanding design risks that were identified should be re-examined to ensure they remain relevant and that any further risks have been identified. Following on from that, the review should consider the following:

- emerging design risks;
- construction/installation risks;
- operational risks;
- maintenance risks;
- disposal risks.

Whilst there may not be mitigations in place for all of these categories of risk at this stage, the specification review should be able to determine additional information that will aid the risk management process over and above what was recorded in the feasibility review. Continually updating the risks in the review process is vital to ensuring that all stakeholders are cognizant of the relevant risks at these project milestones. Additionally, it enables the risk register for the project to be updated with the identification of risk and any mitigation or control measures that have been agreed. It provides, too, a continuity of information through the project to help inform any individual or group (perhaps a supplier) who joins the project later on.

Requirements for the validation of the product which are necessary for it to enter into service should be identified at this stage and the time and cost implications analysed (it is to be noted that some validation tests may be outside of the core competences of the current stakeholder group and may require assistance from specialist agencies with long lead times).

The specification design review is an opportunity to ensure that:

1. Key stakeholders are agreed that the design remains viable or whether it should be halted, revised, or refined.
2. The designer has the right quantity and quality of information to prepare the final technical design.
3. The design remains faithful to the client's statement of requirements.
4. Emerging risks connected to the production and operation of the product are identified and mitigated.
5. Formal approval to move to the next stage of the design process can be agreed.
6. A validation plan exists at this stage.

Formal approval of the design to move on to the next design stage provides the designer with the knowledge that their client has witnessed and agreed all the salient information they (the designer) have provided on the proposed solutions. This is important in providing the designer with the agreement of the specification. Any alterations after this point by the client can be addressed appropriately in terms of further expense to the project and to the designer. Not having this certainty can create friction where late alterations or amendments arise and the reasons for them cannot be readily pinpointed.

The specification design stage should consider that:

- All relevant regulations and standards applicable to the product, its components and materials, and the environment in which it is intended for use are clearly identified. ☑

- This stage is aware of requirements regarding the product's operation in normal, abnormal, and exceptional states. ☑

- Any foreseeable risks to the project and considered mitigation or control measures for these as well as those identified in the pre-construction information have been identified. ☑

- The general principles of prevention continue to be applied during the design process. ☑

- The design continues to meet the original intent outlined in the statement of requirements. ☑

- Any variations to the design requested by the client are accurately documented. ☑

- The core constraints of the project are addressed, ensuring the product will meet the cost/quality/time frame objectives of the client. ☑

- The foundation for a validation plan has been created. ☑

- Acceptance of the design is formally documented. ☑

Full/Technical Design

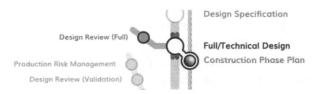

The purpose of the full design (also known as the technical design, final design, or full design intent) is to provide the producer with all the information required from which they can make the product. It should include not only the dimensional specifications (if indeed it has them: software, for example, does not) but also the technical specifications for any material types, component lists, and tolerances that might apply. If the suppliers or producers have been engaged during the previous stages, then the full design should also refer to the manufacturing or installation techniques expected during production. This stage is also an opportunity to ensure that any recommendations or actions recorded in the previous design review have been considered.

The full design would generally consist of technical drawings, lists of parts, and specification sheets. For some projects, the quantity and range of documents could be considerable and therefore the control of them is vital. The competence and structure of the document management system that has been chosen to handle this function must have been defined at the early stages of the project.

During this stage an agreed validation plan is to be finalized, including all the required tests and inspections as well as any prerequisites (such as any necessary plant that is required to conduct the tests, etc.). The validation plan will inform the production programme in terms of key dates and timings. Testing and certification of certain elements may add considerable time to the production process thereby affecting the overall project's delivery schedule.

Design Review—Full

The design review at the end of the design process, as the full design is being prepared to be submitted to the producer, is the last opportunity to ensure that it meets all the objectives of the client's agreed requirements. It is also an opportunity to ensure the design meets the core constraints of the project: can it be made in time; will it be made to budget; and will it be to the expected quality? The capability and capacity of the production stage should also be judged against the requirements of the design, which should now be apparent.

Supplementary to these considerations are whether observable and foreseeable risks have been identified and recorded, and that suitable mitigations or control measures have also been identified as well as naming champions to carry out these measures where possible. The design should be tested against the operational and maintenance requirements—the normal and abnormal states of operation—as well as any exceptional or emergency conditions that may be foreseeable.

The full design review is also where the validation plan can be examined for objectivity and completeness. All of the systems, materials, or components of the design that require validation at the end of the production stage should be identified and the relevant and necessary procedures, certification, or documentation that are required to substantiate the validation process listed against each one. The validation of specialist systems, materials, or components would have been recorded after consultation with the relevant suppliers or manufacturers earlier as part of the additional stakeholder process.

The full/technical design review is an opportunity to ensure that:

1. Key stakeholders are agreed that the design remains viable or whether it should be halted, revised, or refined.
2. The design has reached a suitable level of maturity to be produced.
3. The design remains faithful to the client's statement of requirements.
4. Emerging risks connected to the production and operation of the product are identified and mitigated.
5. The objectives of the validation plan have been agreed by key stakeholders and this is embedded into the production stage.
6. Formal approval to move to the construction/production stage can be agreed.

Construction Phase Plan

The second required document of the Construction (Design and Management) Regulations 2015 is the construction phase plan. This follows the pre-construction information in chronology and is, in essence, a response to it in terms of how risks identified in the pre-construction information will be mitigated or controlled. It should also contain a number of other pieces of information pertinent to the project but should, however, always remain *relevant* and *proportionate* to it.

For the purposes of our methodology for an effective design strategy in projects, we should think of the construction phase plan as the "production stage plan." The contents of this plan will be a valuable asset in not only describing how the production will take place, and how any relevant risks are to be controlled, but also how design amendments are to be managed and by whom. This may be immediately apparent in projects such as those in traditional construction terms, particularly where they are of the design and build type but perhaps not so in other projects. Design amendments can be demanded throughout the production stage and the documented strategy for managing these is important in ensuring the continuity of the design intent (eliminating scope creep); the application of the general principles of prevention; and the prevention of errors.

There is a variety of information that should be considered for inclusion in the plan (see Table 5.10).

Table 5.10 CPP Considerations.

Consideration	Notes
Description	Summary of the project.
	Key dates of the project, such as start date and anticipated or absolute end dates.
Stakeholder details	Names and contact details not only of the production personnel but also any other stakeholders such as client, designers, advisers, and so forth.
Organization	An organizational chart showing the hierarchy of design-making and/or management function.
Safety arrangements	How the production work is to be organized in terms of safety and welfare.
	How specific risks identified in the pre-construction information or risk register are to be controlled.
Design arrangements	The process flow of decision making with respect to design amendments called for during production.
Management	The periodicity, location, and type of meetings (production, design, progress reporting, workgroups, etc.).
	The expected attendees to any particular meeting to ensure it remains quorate.
Training	Any relevant induction training for the production site.
	Any relevant levels of expertise required for the production process.
Production	Description of any relevant production techniques or processes that involve particular risks or require elevated levels of control or management.
Engagement	How production personnel will be engaged with the production process: feedback, toolbox talks, work groups, and so forth.
Welfare	What welfare provisions are to be available at the production location.
	Any specific provisions required for risks identified in the pre-construction information.
Emergencies	Any relevant provisions for dealing with emergencies.
	The hierarchy of notifying senior management of emergencies and the process for doing so.

Each individual project will require a different approach to the type and level of information contained in the plan, but it should always remain relevant. Remember also that this information is pertinent to the production stage *only* and should remain "live" (i.e., able to be reviewed and updated throughout) during the production stage. In large, complex projects it is entirely possible for the plan to be revised several times, perhaps as a result of design amendments or the arising of risks that had not hitherto been foreseeable. It is possible that a change in specification of components or materials, perhaps where the original choice becomes unavailable, causes a change in the production techniques required.

The plan should remain visible to all concerned with the production stage and should be reviewed regularly to ensure its continued validity. This may be done at meetings throughout the production stage with relevant stakeholders.

The full/technical design stage should consider that:

- A robust design review takes place that identifies that the final design: ☑

  ○ meets the statement of requirements; ☑

  ○ reduces all risks to acceptable levels; ☑

  ○ can be produced; ☑

  ○ can be operated and maintained safely. ☑

- the construction phase plan demonstrates how risks will be managed throughout the production stage: ☑

- the management of the design objective is maintained throughout the production stage: ☑

- there is full and final agreement on progressing to the production stage: ☑

- the risk register is updated to help inform the production stage plan: ☑

- there is formal acceptance of the finalized validation plan: ☑

- training and competencies have been considered: ☑

Production

The design process continues to form an important part of the production stage. This is, in part, due to emerging risks that may or may not have been previously identified and recorded either in the construction phase plan, the risk register, or both. There are several examples of emerging risks that might affect our case studies as they enter the production stage (see Table 5.11).

Table 5.11 Possible Emerging Risk During Production.

Case study	Examples of possible emerging risks
Nuclear power plant	The supply of a printed circuit board for the reactor monitoring equipment fails and the system requires redesign using an alternative, potentially extending the timescale of the build.
Office block	The discovery of important archaeological remains requires a variation of the basement level layout and supporting structure.
Warship	The factory acceptance test of a new engine reveals that the fuel flow is actually insufficient resulting in the pipework having to be re-engineered.
Home printer	The moulding technique for the print head shuttle produces a built-in weakness causing early failure of the component resulting in a redesign of the moulding tool and specification of material type.
Motor car	Legislation demands the production line must reduce its carbon footprint resulting in a redesign of manufacturing tools, material types, and production techniques, which must be integrated with current production output.

All of the above examples may cause a significant redesign but there are many other examples where small changes are required, perhaps due to supply chain difficulties, emerging geo-political situations, labour shortages, or even design errors that have gone unnoticed in the process so far. Errors are always possible, irrespective of the project resources devoted to their elimination. It is therefore important to have well-documented control and management of the design process, as well as reliable communication channels set up in order to effectively deal with errors and emerging risks. Preventing errors in the first place is a good thing; also being able to effectively deal with them if they arise is even better.

Production Risk Management

It is important that the risk register is reviewed and updated as required through-out the production stage. This is to not only ensure that the management of risk is maintained during the production phase, but also to inform the health and safety file and/or safety case, which will be required prior to the product entering service. The purpose of these documents is to inform the owner and/or end user of any residual risks or hazards that remain, as well as to provide vital safety information for those who have to maintain, repair, or dispose of the product. This is the trail of information, or "golden thread," running throughout the course of the product's life cycle. It may also help to inform any necessary training, instruction, or supervision of the product that is required and, of course, any CE marking that should be completed.

For complex engineering designs, the requirement for a safety case is to demon-strate that the design is safe for all its intended uses. A safety case is a body of evi-dence that identifies the hazards and risks and the associated mitigations and possibly operating restrictions that, if followed, will ensure safe operation. As such, it differs little from the requirements of both the health and safety file and the techni-cal file for CE marking, although will by necessity be somewhat more technically in-depth.

Design Review—Validation

At the end of the production stage, and before the product enters service, the final design review enables the project team to assess the following.

- Has the product met with the statement of requirements?
- Has the product been delivered to the agreed level of quality?
- Does the product meet with all regulatory requirements in terms of use, mainte-nance, and disposal?
- Have all the systems that require formal assessment prior to use been validated appropriately and is the documentation in order and available?
- Have all risks been reduced to a level as low as reasonably practicable where pos-sible and have residual risks been identified and documented?
- Have operating restrictions or limitations been identified and are any mitigations appropriate for the end user's understanding and capabilities?

These are crucial questions to answer before the product enters service. They dem-onstrate that all of the stakeholders involved in the project have executed their responsibilities correctly, professionally, and within the requirements of any regula-tions that may apply to either them, their work, or the product itself.

In complex projects, like our nuclear power station example, where elements of the project have to be brought online consecutively, this review should take place at the completion of each of these elements. This will ensure that the relevant operational checks have been performed prior to the next element beginning. This is vital to ensuring the ongoing safety of the project and also provides a valuable documentary trail of decisions made and information garnered.

In terms of residual risks, the validation design review is an opportunity to review these to ensure all the relevant information about them has been collated for inclusion in any relevant safety documentation.

Acceptance/Handover

Once the validation design review is complete, and the stakeholders are agreed that the product can enter service, a formal acceptance should take place. This will be from the client to the controlling authority for the design *at the point that production ceases*. This may be the producer, the designer, the project manager, or some other entity and will have been decided and documented as part of the management structure at the beginning of the project. This formal acceptance again demonstrates that all parties have concluded their responsibilities appropriately, as well as documenting that the client has accepted the final output of the project thereby releasing the producer from their involvement in the design process. Where the producer is contractually obliged to continue their involvement with the product whilst it is in use, perhaps for the purposes of setting-to-work, training, or snagging, the formal acceptance also provides a reference point for the date from which this activity commences.

The acceptance document package, passed to the client upon formal acceptance into service, might contain the following.

- The risk register, detailing risks mitigated, controlled, and outstanding.
- Details of any limitations of operation.
- Details of any systems that are not in service, perhaps where the client is to integrate them at a later date (lighting or CCTV systems, for example).
- Approval certificates for structures, materials, or components.
- Test certificates.
- Maintenance requirements, procedures, and periodicity, and/or manufacturer's guidance notes.
- Details of any hazardous materials, their location, and any special precautions or handling techniques.
- Drawings, plans, schematics, and general arrangements.
- Safety case(s).
- Limitations of operation.

The formal acceptance should also identify any training requirements for the use, maintenance, and disposal of the product. This may be in the form perhaps of specialist training that is required to a manufacturer's stipulation or a general or accredited level of qualification.

Health and Safety File

The health and safety file is the third and final documentary requirement of the Construction (Design and Management) Regulations 2015. In Appendix 4 of the Regulations, it states that the file "must contain information about the current project likely to be needed to ensure health and safety during any subsequent work, such as maintenance, cleaning, refurbishment or demolition." From this, we can see that much of the information required for the file has already been included in the acceptance documentation described above. However, it should be appreciated that the reason for the health and safety file is different to that of the acceptance documentation. The acceptance documentation is for the *client*, and is intended to provide them with all the relevant information, including that involving safety: firstly, to demonstrate that they have the product they have paid for; secondly, that the product meets or exceeds regulatory requirements; and thirdly, what the client can expect in terms of operability.

The health and safety file is for the *end user(s)* and is intended to cover all aspects of safety-related operation of the product throughout its use, maintenance, and ultimate disposal. To elucidate this difference, let us consider an electrical test certificate issued for an office block. The installation and testing of electrical wiring in commercial (and domestic) premises in the UK must meet the requirements of BS 7671:2018—Requirements for Electrical Installations, IET Wiring Regulations (IET, 2018) and is overseen by a number of electrical inspection bodies regulated by the government. Commercial electrical systems cannot be used before they have been appropriately tested and certificated.

The purpose of the certificate in the acceptance documentation is to prove to the client that the relevant tests have been completed and that the electrical system is to the standard(s) required. It enables the premises to be commissioned into service, all the while providing evidence of the client's responsibility for ensuring the safety of the end user. Its inclusion in the health and safety file is to indicate to any subsequent electrician working on the system how the system is connected and the type and rating of components within it. This ensures the safety not only of any individual working on the system but also that of those around them and the building itself.

A list of the other information that should be considered for inclusion in the health and safety file is identified in Appendix 4 of the Construction (Design and Management) Regulations 2015 (see Table 5.12).

Table 5.12 CDM 2015 Appendix 4.

---

a) a brief description of the work carried out;

b) any hazards that have not been eliminated through the design and construction processes, and how they have been addressed (e.g., surveys or other information concerning asbestos or contaminated land);

c) key structural principles (e.g., bracing, sources of substantial stored energy—including pre- or post-tensioned members) and safe working loads for floors and roofs;

d) hazardous materials used (e.g., lead paints and special coatings);

e) information regarding the removal or dismantling of installed plant and equipment (e.g., any special arrangements for lifting such equipment);

f) health and safety information about equipment provided for cleaning or maintaining the structure;

g) the nature, location and markings of significant services, including underground cables; gas supply equipment; fire-fighting services etc.;

h) information and as-built drawings of the building, its plant and equipment (e.g., the means of safe access to and from service voids and fire doors).

Important points to remember about the health and safety file are that:

- the level of detail in the information is proportionate to the risks involved;
- the information is provided in a clear, concise, and convenient form;
- the information is accessible and easily understood;
- the information is amended throughout the lifetime of the product to reflect any changes or additions to it;
- the health and safety file remains with the product for the entirety of its operational life and is made available to anyone who has reason to view its contents.

The responsibility for ensuring the creation of the health and safety file under the CDM regulations rests with the client, but it is the principal designer who must begin assembling its contents during the design process. At the end of the project either they, or the principal contractor (producer), must complete the file for submission to the client. However the file is generated, it can be seen from the list of relevant information that should be included in the file that it is vital to anyone, during the lifetime of the product, who is engaged to amend, redesign, dispose of, or add to the product or any component of it.

The relevance of creating and disseminating the file properly should not be underestimated. The entire design process, from start to end, is guided by having the relevant information to hand at the appropriate time and managing that information in a controlled way. The output of the design process—a finished product—relies equally on information being available in order that it can be used or operated effectively and safely. This can only be done if its hazards and limitations of use are identified and recorded. This is, in effect, the idea behind the "golden thread," which calls for relevant safety information to be made available and managed accordingly throughout a product's life cycle. It is a venerable initiative. However, the reasons and tools for creating a "sphere of safety" around our designed products have long been in existence and it is our legal duty to abide by them and our moral duty to implement them.

The production stage should consider that:

- Any modifications to the design have been clearly captured and identi- ☑ fied, the impact measured, and the risk register updated.

- A complete risk register is provided on completion of acceptance. ☑

- A validation design review is completed to consider the output against ☑ the original design intent, as well as to examine any residual risks.

- The contents of the acceptance documentation meet with any specific ☑ requirements.

- Any specific training requirements are documented. ☑

- A health and safety file is created and delivered. ☑

- Where there is requirement for an additional safety case, this is devel- ☑ oped and issued.

In Service

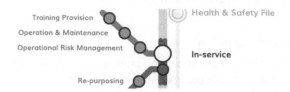

Training Provision
Operation & Maintenance
Operational Risk Management

Health & Safety File

In-service

Re-purposing

Although the design process technically finished on completion of product valida-tion and acceptance, there may be a requirement to start new design processes as a result of an emerging need to change the product to meet new requirements. These may be driven by such things as:

- a change in legislation;
- entering a new market that has different regulations or a different interpretation of existing regulations;
- the requirement for additional functionality;
- a result of historic failure data or warranty issues;
- improvements in maintenance techniques;
- the availability of new lighter, cheaper, or more sustainable materials.

Whatever the reason for updating or redesigning the product, the design process must start from the beginning (initiating need).

Whilst a product is in service, there remains a need to capture information on how it is used or operated, whether there are any issues during its maintenance or repair, and whether there are any significant failures. We discussed at the beginning of this book the various influences that affect design. To recap, they included:

- The initiating need of the client and the statement of requirements.
- Safety-related matters to do with the pre-construction or pre-production information.
- Relevant risks to the design including financial, reputational, qualitative, and so on.
- Previous experience of similar products or operational environments.
- External influences: social, political, industrial, technological, marketplace, and so forth.

Those individuals and teams involved in operations, quality control, health and safety, engineering, and human resources can all assist in collecting and managing data on these matters. This will not only aid any design amendments during the in-service period but also any future designs of similar products. Any information perti-nent to the safety of the product, or those who interact with it, should be appended to the health and safety file. This will further help any future amendments to the product at the initial stages when pre-construction information, or pre-production information, will be required.

The reasons for a design failing during the in-service period are many and varied. If the design process was well executed and the stakeholders—and most importantly the designer— all acted competently then hopefully it will not be for want of dili-gence. External forces and factors may, however, emerge and conspire to effect a requirement for change (see Table 5.13).

Table 5.13 Examples of Possible Emerging Risks.

Case study	Examples of possible emerging risks
Nuclear power plant	A modified design of the flood defences is required due to a recent catastrophic event involving a similar power plant.
Office block	The need to separate pedestrian traffic flows in elevators, stairwells, and corridors due to a pandemic.
Warship	The discovery by the intelligence services of new enemy detection systems requiring the need to improve or alter the ship's defensive capabilities.
Home printer	The banning of an ink ingredient in the destination market due to environmental concerns causing a reworking of the printer head.
Motor car	The merger of the client with another manufacturer resulting in the re-specification of a number of components due to cost benefits.

Whatever factor becomes the driver for design change in service, it follows that the process to bring about that change is based on the sound methodology we have described and is supported by relevant, practical, and available information. Updating that information throughout a product's life cycle can only be a pragmatic and sensible approach. As we have seen, the design process is supported by information contained in such things as the pre-construction information, the construction phase plan, and the health and safety file. Other documents such as safety cases and technical files all fully support this "cradle to grave" assemblage of information.

This supports the principle of the "golden thread" that runs through a product's entire life cycle: of informing every stage with relevant safety information in order to promote better design in the future and to reduce the risk of harm. It could also be that this information could help to inform the repurposing or recycling of the product in order to make either the product itself more sustainable, or those that follow it.

It is irrelevant how the information is presented: what is important is that it *is available* and *properly updated*. This requires the intervention of competent people throughout the life cycle of the product assuming the responsibility for improving safety for everyone.

Risk Management in Service

Once in operational service, the risks associated with the product will be influenced by:

- changes in legislation;
- issues identified through failure rate data;
- changes in operational environment;
- technological and scientific advancements;
- obsolescence.

These are emergent risks and can often be managed throughout the product's service life. The management of risks during this period are incumbent on the operator under the relevant health and safety legislation. This may include any or all of the following.

- Documenting an assessment of risk.
- Detailing the policies, processes, and procedures for operating the product.
- Providing suitable information, instruction, and training.
- Supervising the use of the product.
- Providing guarding or demarcating work areas.
- Ensuring the use of personal protective equipment.

It may be seen from this list that many of the situations that could require the use of control measures might, with suitable reference to the general principles of prevention, be avoided during the design stage. Should any of the risks during the operational stage, whether emergent or not, become intolerable for any reason, there may well be the need to mitigate them through design-related measures such as mid-life design updates, repurposing, and/or early retirement/disposal of the product.

Training Provision

The level and type of training required for any product will have been established during the design process and will remain a residual risk in terms of operation of the product throughout its lifetime. This is because such training will have to be continually conducted throughout this time and is therefore subject to such influences as competence of the trainer, validity over time, and format. Complex face-to-face training for a product using a specific delivery mechanism is likely to suffer from obsolescence over time. This may be because of being able to provide competent training personnel throughout the duration, or the possibility of the delivery mechanism becoming redundant, or even changes in training methods or general levels of competency among operators.

Devices with menu-driven electronics can easily be loaded with help text so long as this remains valid for any future updates. The question of how updates are delivered is also a risk: if via a hard mechanism like a DVD or plug-in, then how is this to be administered; if via the internet then what happens if the delivery is missed or the device is able to be hacked? Pictograms and signage are another effective way of providing training that has no time limit so long as they remain valid, readily identifiable, and translate appropriately into different languages and cultures.

Consider a production machine in a factory. We mentioned previously about control interfaces needing to consider people who may have problems with perception, that is, they may have difficulty with seeing, reading, or cognition. Using elaborate ideograms on the control buttons of our production machine may well fit the designer's aesthetic or the latest design trends, but will it need reams of paper to explain it? Simple "up/down," "yes/no" buttons may not be à la mode but will they require less training? If the machine is to be operated by skilled personnel, what level of skill do we specify as being required to ensure that (1) our ideograms are understood, and (2) that mistakes in pressing the wrong button are reduced as much as possible? The foreseeable risk of what happens *if* the wrong button is pressed is another factor for the designer to consider.

Training can often be an afterthought of the design process, and sometimes we can be guilty of assuming that operators will understand the product simply because we have designed something that is similar to what is already available. This can be perfectly adequate if the product does indeed truly depend on "life skills," such as driving, for example. However, the level of training should have been considered at the outset of the project (see Table 5.14) and the risks of inadequate training, or misoperation of the product due to inadequate training suitably recorded in the risk register.

Table 5.14   Training Considerations.

Stage	Possible considerations
Initiating need	This will be the creation of a product for which training will be required. This stage will interact with the design process of the product itself insomuch as they will inform each other about the type and level of training that is required and appropriate.
Business case	Consider if the training can be developed in-house or whether it requires external intervention. Establish any early risks with how the training will impact the project or the product and the level of adequacy required.
Statement of requirements	Clearly identify the requirements of the training provision whilst ensuring these inform but do not control the training design process.
Initiating the design process	Establish the management structure, document management system, and key stakeholders required for the process.
Concept design	Consider all possible ways of delivering the training programme, including the novel or unorthodox. Consider the general principles of prevention in relation to preventing risks with the *delivery* of the training.
Additional stakeholder engagement	Identify additional stakeholders who may have an influence on, or be influenced by, the training provision. This might be HR departments, training providers, service departments, retail and wholesale hubs, field engineers, and so forth.
Specification design	Develop the concepts into a defined design specification, taking note of all external influences on how the training provision should be delivered. Ensure the specification takes account of the original needs and requirements of the client (the development of the product itself).
Full/technical design	Review the training provision to ensure all the relevant factors have been accounted for and any risks with its creation and/or delivery have been identified and suitably controlled.
Production stage	Produce the training provision, whether this is an online course, document package, manual, and so forth. Ensure the provision remains true to the original intent and specification.
In use	Monitor the training through the lifetime of the product to ensure it remains valid, especially if there are variations or extensions of the product.
	If the product is repurposed, the training will have to be thoroughly reviewed to ensure that it includes any revisions of the operational parameters.

Operation and Maintenance

The operational and maintenance parameters of the product will have been defined during the design process, having been suitably informed by external influences and additional stakeholders. For simple, mass-produced products, it may be accepted that the design intent and operability of the product is unlikely to change in its lifetime. For complex products, however, it can be easily presumed that the operation and maintenance of the product might alter over time. Not due to the design, nor the intent of the design, but because of the *operational environment*. This includes the operators and maintainers themselves.

A simple example would be a factory building designed for light manufacturing which, over time, begins to change to heavy manufacturing. The considerations for a building containing heavy machinery—floor strength, wall thickness, soundproofing, power supply, and so forth—would be quite different from the original intent of the building and, therefore, would require an effective design strategy to be re-implemented.

But operational changes over time can also be as a result of human interaction. Over time, changes in management and supervision can affect the way that a product is operated, sometimes due to the rate at which operators are replaced. The rate of staff turnover, or "churn," can impact greatly on product operation as the original training provision becomes an ever-more distant memory, especially if the churn includes supervisory and training roles. Once again, understanding the client's business model as part of the design process can highlight staff turnover as a risk to the effective continued operation of the product, and can be mitigated through not only design measures but also in the type and level of training provision. Highlighting critical operational points on the product itself is one example of how such mitigation might be put into place.

Maintenance, too, is a function that, over time, can alter from the prescribed specification. Motor cars require servicing at prescribed intervals, for example, but it is often the case that as a car ages, and the vehicle becomes less valuable, routine servicing does not take place at the appropriate times. Modern cars have overcome this problem to some extent by having onboard sensors that monitor the time and mileage of the vehicles, and indicate that a service is required via the dashboard. Some cars can even monitor the way they have been driven in order to determine when a service is due based on the stresses they have undergone due to the operator's driving style. This improvement in car design was in some way led by the use of telemetrics on racing cars (particularly in Formula One) where data from many sensors on the cars are processed remotely to identify operational issues during a race. This allows engineers in the pits to call in the cars for repairs or modifications before the problem becomes serious, or possibly even obvious to the driver.

Simpler maintenance issues might arise from, say, placing a maintenance access panel at the rear of a machine that is then installed too close to a wall to allow the panel to be opened fully. This may cause the maintenance engineer to find an alternative access point, which will undoubtedly be a less safe solution. It should always be borne in mind that maintainers are engineers and as such their job is to solve problems. If the problem is one of access their psychology is to find a way to resolve the issue. Preventing any form of improper access; highlighting the amount of access space around the machine; issuing well-defined training provisions; and ensuring the delivery of a complete and well-documented health and safety file are ways that this particular issue could be controlled.

Repurposing

Repurposing is a clear example of changing the original design objective. Ships that become hotels; power stations that become business centres; office blocks that become social housing—all of these have moved some distance from their original statements of requirements.

That we must re-implement an effective design strategy from the moment that the repurposing of the product is first mooted is without question. And that the strategy should be exactly as we have described in this chapter is also true, albeit that some elements may be briefer and others more involved. What is also clear is that the business case and early conceptual stages will be influenced by the information contained in the health and safety file from the original incarnation of the product. Indeed, it is important for any pre-construction information to consider the information contained in this file.

The in service stage should consider that:

- The provision of any training, instruction, supervision, or signage is ☑ maintained throughout the lifetime of the product.

- There is an ongoing assessment of operational and emerging risk. ☑

- The onward submission of safety information is related in cases of the ☑ product's sale, disposal, or repurposing.

- Operation and maintenance of the product remains within any limita- ☑ tions of use and the designed objectives.

Disposal

Firstly, we need to understand that disposal in this sense means the physical intervention of preventing the product from operating in its intended form. This includes:

- demolition;
- deconstruction;
- recycling;
- destruction.

The mere transference of ownership—the "disposal" of the product as an asset—only concerns us in respect of the documentation that must accompany the transfer. That is to say such things as the health and safety file, the technical file, the safety case, operating instructions, and maintenance records and requirements.

Where products are to be simply thrown away at the end of their designed life, such as a home printer, then the implication of that form of disposal will have been covered in the specification design stage, having been suitably informed by the statement of requirements. This will have included any regulatory requirements too, such as the Waste Electrical and Electronic Equipment Regulations 2013 (WEEE) (Environment Agency, 2013). Although these regulations have been in place for some time, and of course could be amended during the anticipated lifetime of a product being designed today, the mere act of designing a product that is to be simply thrown away should instil in the designer a moral duty to prevent unnecessary environmental impact.

In the case of larger products that need dismantling or demolition, there are two important points to consider. Firstly, although the method of dismantling or demolition may not be a consideration for the designer, the presence of elements within the design that may cause an issue certainly is. Once again, the creation of an appropriate health and safety file for the product, and its continued availability with the product throughout its operational life, cannot be underestimated. The file will detail the presence and location of hazardous materials and structural elements that could cause harm during the dismantling or demolition stage.

An extreme example is that of asbestos. Widely used in construction until its complete ban in the UK in 1999, asbestos is perhaps the most hazardous of all nonradioactive materials and is still responsible for circa 5,000 deaths in the UK each year. The identification of the presence and location of asbestos-containing materials in the health and safety file for buildings would have gone some way to preventing both harmful exposure to it during subsequent intrusive investigations prior to its removal, as well as the enormous costs this has had in both monetary and human terms. The requirement for creating a health and safety file came about in the construction regulations introduced in 1994.

Secondly, a demolition, deconstruction, or dismantling project is, in itself, a design process, albeit in reverse. It will, therefore, require an effective design strategy in line with that which we have discussed throughout this book. As with the strategy for training and repurposing, the strategy can obviously be tailored to the precise requirements and implications of the project itself. The important point is that all the points raised by the strategy are considered and managed appropriately and effectively. The one key difference is that the documentary output of the project will not be the health and safety file as before, but a disposal risk assessment, which may have to include an environmental impact assessment.

Disposal Risk Assessment

One of the purposes of the health and safety file is to inform those who have to dismantle or dispose of the product at the end of its serviceable life of any hazards or risks that exist. This might be the location and type of any hazardous materials, high-energy systems—such as power or hydraulics, or stored energy systems such as pre- or post-stressed concrete members. Much the same information will also be contained in a safety case for a complex engineering product. The regulatory requirement for health and safety is evident: It is to provide the necessary information to prevent harm and therefore its creation and completion is not just a legal requirement but a moral duty.

The duty of anyone completing a risk assessment at the point of disposal is clearly laid out in general health and safety regulation and their ability to do so is greatly aided by knowledge that stems from the time of the original design and construction of the product. Under the Construction (Design and Management) Regulations 2015, demolition is one of the activities considered "construction" and, therefore, the whole process of providing pre-construction information and a construction phase plan, as well as ensuring the provision and management of a safe working environment will be legally required. Consider how much easier—and safer—this would be with adequate and complete reference information.

Additionally, information regarding hazards and residual risks for a product will be invaluable for any environmental impact assessment that has to be completed when disposing of it. It will also inform any repurposing efforts where the product is absorbed into the circular environment to be reused or recycled. The provision of the right information throughout a product's life cycle is simply about providing it to the right people at the right time.

The disposal stage should consider that:

- The risks of disposal have been considered and mitigated where possible.

- Any regulations surrounding disposal are identified.

Bibliography

APM. (2019). *APM Body of Knowledge* (7th edition ed.). (APM, Ed.) Buckinghamshire: APM. doi: ISBN: 978-1-903494-82-0

Autodesk Building Industry Solutions. (2002). *Building Information Modeling*. San Rafael USA: Autodesk inc.

BSI. (2012). PAS 99:2012 - *Specification of common management system requirements as a framework for integration*. London: BSI Standards Limited. doi: ISBN 978 0 580 76869 9

BSI. (2018). *BS EN IEC 60812:2018 - Failure modes and effects analysis (FMEA and FMECA)*. doi: ISBN: 978 0 580 87537 3

BSI. (2018, February 14). ISO 31000:2018 - Risk management. Guidelines. BSI. doi: ISBN 978 0 580 88518 1

BSI. (n.d.). BS EN 60529:1992 - Degrees of protection provided by enclosures (IP Code). *BS EN 60529:1992*. London. doi: ISBN 978 0 539 03833 0

BSI. (n.d.). PAS 99 *Integrated Management Systems*. Retrieved November 19, 2020, from BSIGroup: https://www.bsigroup.com/en-GB/pas-99-integrated-management/.

Cabinet Office. (2011). *Government Construction Strategy*. Cabinet Office. London: Cabinet Office.

CQI. (2016). MSS 1000:2014 - *Management System Specification and Guidance*. Chartered Quality Institute. Retrieved from https://www.researchgate.net/publication/293314132_MSS_10002014_Management_System_Standard

Edwards v National Coal Board, 1 All ER 743 ((CA) 1949).

Environment Agency. (2013). *Waste Electrical and Electronic Equipment Regulations (WEEE)*. London: HMSO.

Ericson, C. (1999). Fault Tree Analysis - A History. *Proceedings of the 17th International System Safety Conference*. Retrieved May 5, 2012, from http://www.fault-tree.net/papers/ericson-fta-history.pdf

GIRI. (2019, June). *Strategy for change*. Retrieved 2020, from Get it Right Initiative: https://getitright.uk.com/live/files/reports/2-giri-strategyforchange-revision-june2019-589.pdf

Haddon-Cave, C. (2009). *The Nimrod Review*. London: The Stationary Office. Retrieved November 14th, 2010

HSE. (1974). *Health And Safety At Work Etc. Act 1974*. London: Crown copyright. Retrieved October 2020, from http://www.legislation.gov.uk/ukpga/1974/37.

An Effective Strategy for Safe Design in Engineering and Construction, First Edition.
David England & Dr Andy Painting.
© 2022 John Wiley & Sons Ltd. Published 2022 by John Wiley & Sons Ltd.

HSE. (2013). HSG65 - *Managing for health and safety* (Third ed.). HSE. doi: ISBN 978 0 7176 6456 6

HSE. (2015, April). L153 - Managing health and safety in construction. HSE Books.

HSE. (2019). *Construction statistics in Great Britain.* HSE.

HSE. (n.d.). *The Workplace (Health, Safety and Welfare) Regulations 1992.* Crown

IET. (2018). BS 7671:2018 - *Requirements for Electrical Installations. IET Wiring Regulations* (18th ed.). London: Institute of Engineering and Technology. doi: ISBN: 978 1 785 61170 4

Institute of Risk Management. (NK). *Risk appetite and tolerance.* Retrieved from Institute of Risk Management: https://www.theirm.org/what-we-say/thought-leadership/risk-appetite-and-tolerance.

ISO. (2015). *ISO 9001:2015 - Quality Management Systems - Requirements.* Retrieved November 2020

Luft, J., & Ingham, H. (1955). The Johari window, a graphic model of interpersonal awareness. *Proceedings of the Western Training Laboratory in Group Development.* Los Angeles: University of California.

Marvin Rausand, A. H. (2004). *System Reliability Theory – Models, Statistical Methods, and Applications* (2nd ed.). Wiley. doi: ISBN: 978-0-471-47133-2

Ministry of Housing, Communities and Local Government. (2018). *Building a Safer Future - Independent Review of Building.* Secretary of State for Housing, Communities and Local Government. London: Crown Copyright. doi: ISBN 978-1-5286-0293-8

NASA. (2012, May 4). PD–AP–1307 - Failure Modes, Effects, and Criticality Analysis (FMECA). Retrieved from http://engineer.jpl.nasa.gov/practices.html

Reason, J. (1990, April 12). The contribution of latent human failures to the breakdown of complex systems. 327(1241), 475-484. doi: https://doi.org/10.1098/rstb.1990.0090

Reason, J. (2000, March 18). Human error: models and management. *BMJ, 320,* 768-770.

RIBA. (n.d.). RIBA Plan of Work 2020 Overview. London: RIBA. Retrieved September 13, 2020, from https://www.architecture.com/-/media/GatherContent/Test-resources-page/Additional-Documents/2020RIBAPlanofWorkoverviewpdf.pdf

Springer Series in Reliability Engineering. (2010). Probabilistic Safety Assessment. *Reliability and Safety Engineering, 0,* pp. 323-369. doi: https://doi.org/10.1007/978-1-84996-232-2_9.

Taleb, N. N. (2007). *The Black Swan: The Impact of the Highly Improbable.* The Random House Publishing Group

The W. Edwards Deming Institute. (2020). *PDSA Cycle.* Retrieved from The Demming Institute: https://deming.org/explore/pdsa/

UK Statutory Instruments. (2020). Town and Country Planning (Environmental Impact Assessment) Regulations 2017. London: legislation.gov.uk. Retrieved from Legislation.gov.uk: https://www.legislation.gov.uk/uksi/2017/571/made

Index

An Effective Strategy for Safe Design in Engineering and Construction, First Edition.
David England & Dr Andy Painting.
© 2022 John Wiley & Sons Ltd. Published 2022 by John Wiley & Sons Ltd.